Reliability-Based Mechanical Design Volume 1

Component under Static Load

Synthesis Lectures on Mechanical Engineering

Synthesis Lectures on Mechanical Engineering series publishes 60–150 page publications pertaining to this diverse discipline of mechanical engineering. The series presents Lectures written for an audience of researchers, industry engineers, undergraduate and graduate students.

Additional Synthesis series will be developed covering key areas within mechanical engineering.

Solving Practical Engineering Mechanics Problems: Kinematics
Sayavur I. Bakhtiyarov
2018

C Programming and Numerical Analysis: An Introduction
Seiichi Nomura
2018

Mathematical Magnetohydrodynamics
Nikolas Xiros
2018

Design Engineering Journey
Ramana M. Pidaparti
2018

Introduction to Kinematics and Dynamics of Machinery
Cho W. S. To
2017

Microcontroller Education: Do it Yourself, Reinvent the Wheel, Code to Learn
Dimosthenis E. Bolanakis
2017

Solving Practical Engineering Mechanics Problems: Statics
Sayavur I. Bakhtiyarov
2017

Unmanned Aircraft Design: A Review of Fundamentals
Mohammad Sadraey
2017

Introduction to Refrigeration and Air Conditioning Systems: Theory and Applications
Allan Kirkpatrick
2017

Resistance Spot Welding: Fundamentals and Applications for the Automotive Industry
Menachem Kimchi and David H. Phillips
2017

MEMS Barometers Toward Vertical Position Detecton: Background Theory, System Prototyping, and Measurement Analysis
Dimosthenis E. Bolanakis
2017

Engineering Finite Element Analysis
Ramana M. Pidaparti
2017

Reliability-Based Mechanical Design, Volume 1: Component under Static Load
Xiaobin Le

ISBN: 978-3-031-79636-4   paperback
ISBN: 978-3-031-79637-1   ebook
ISBN: 978-3-031-79638-8   hardcover

DOI 10.1007/978-3-031-79637-1

A Publication in the Springer series
SYNTHESIS LECTURES ON MECHANICAL ENGINEERING

Lecture #20
Series ISSN
Print 2573-3168   Electronic 2573-3176

Reliability-Based Mechanical Design Volume 1

Component under Static Load

Xiaobin Le
Wentworth Institute of Technology

SYNTHESIS LECTURES ON MECHANICAL ENGINEERING #20

ABSTRACT

A component will not be reliable unless it is designed with required reliability. *Reliability-Based Mechanical Design* uses the reliability to link all design parameters of a component together to form a limit state function for mechanical design. This design methodology uses the reliability to replace the factor of safety as a measure of the safe status of a component. The goal of this methodology is to design a mechanical component with required reliability and at the same time, quantitatively indicates the failure percentage of the component. *Reliability-Based Mechanical Design* consists of two separate books: *Volume 1: Component under Static Load*, and *Volume 2: Component under Cyclic Load and Dimension Design with Required Reliability*.

This book is *Reliability-Based Mechanical Design, Volume 1: Component under Static Load*. It begins with a brief discussion on the engineering design process and the fundamental reliability mathematics. Then, the book presents several computational methods for calculating the reliability of a component under loads when its limit state function is established. Finally, the book presents how to establish the limit state functions of a component under static load and furthermore how to calculate the reliability of typical components under simple typical static load and combined static loads. Now, we do know the reliability of a component under static load and can quantitively specify the failure percentage of a component under static load.

The book presents many examples for each topic and provides a wide selection of exercise problems at the end of each chapter. This book is written as a textbook for junior mechanical engineering students after they study the course of Mechanics of Materials. This book is also a good reference book for design engineers and presents design check methods in such sufficient detail that those methods are readily used in the design check of a component under static load.

KEYWORDS

reliability, reliability-based design, mechanical component, mechanical design, computational method, numerical simulation, static load, limit state function, failure, safety, probability

To my lovely wife, Suyan Zou,
and to my wonderful sons, Zelong and Linglong

Contents

Preface

Reliability-Based Mechanical Design consists of two separate books: *Volume 1: Component under Static Load*, and *Volume 2: Component under Cyclic Load and Dimension Design with Required Reliability*.

Volume 1 consists of four chapters and Appendix A. They are:

- Chapter 1: Introduction to Reliability in Mechanical Design;

- Chapter 2: Fundamental Reliability Mathematics;

- Chapter 3: Computational Methods for the Reliability of a Component;

- Chapter 4: Reliability of a Component under Static Load, and,

- Appendix A: Samples of MATLAB Programs.

Volume 2 consists of three chapters and two appendixes. They are:

- Chapter 1: Introduction and Cyclic Loading Spectrum;

- Chapter 2: Reliability of a Component under Cyclic Load;

- Chapter 3: The Dimension of a Component with Required Reliability;

- Appendix A: Three Computational Methods for the Reliability of a Component; and,

- Appendix B: Samples of MATLAB Programs.

The first book discusses fundamental concepts for implementing reliability in mechanical design and the reliability of a component under static load. The second book presents more advanced topics, including the reliability of a component under cyclic load and the dimension design with required reliability.

Why does a component fail even the factor of safety of a component is more than the required factor of safety, for example, 2.5? If a component will fail, what is its possible percentage of the failure? This book presents how to determine the reliability, and quantitively predict the percentage of the failures of a component under static load. Therefore, we can provide the reliability and also indicate a possible percentage of failure of a component under static load.

This book is written as a textbook and is based on a series of lecture notes of an elective course for junior mechanical students. Every topic is discussed in sufficient detail and demonstrated by many examples so students or design engineers can readily use them in mechanical design check. At the end of each chapter, there is a wide selection of exercises. This book can also be used as a reference book for design engineers.

This book consists of four chapters and Appendix A. A concise summary of each chapter are as follows.

- Chapter 1: Introduction to Reliability in Mechanical Design

  This chapter serves as an introduction and will discuss the engineering design process, failures, and uncertainty in engineering design, reliability definition, and history.

- Chapter 2: Fundamental Reliability Mathematics

  This chapter discusses the fundamental concepts and definitions of probabilistic theory for the preparation of their implementation for mechanical design. This chapter enables a person without any knowledge of probability theory to use this book to conduct the reliability-based mechanical design.

- Chapter 3: Computational Methods of the Reliability of a Component

  This chapter discusses several computational methods of the reliability of a component when the limit state function of a component under load is established. Those methods include the interference method, the First-Order Second-Moment (FOSM) method, the Hasoder-Lind (H-L) method, the Rachwitz-Fiessler (R-F) method, and the Monte Carlo method.

- Chapter 4: Reliability of a Component under Static Load

  This chapter presents typical limit state functions of a component under each typical static load and combined load, and further demonstrate how to calculate the reliability of components under any type of static loads. Five typical component cases presented in this chapter include a bar under axial static load, a pin under direct shear static load, a shaft under static torsion, a beam under static bending moment, and a component under combined static loads.

- Appendix A: Samples of three MATLAB Programs

  Appendix A provides three MATLAB programs as references for calculating the reliability of a component under static load. These three samples of MATLAB programs include one for the Hasoder-Lind (H-L) method, one for the Rachwitz-Fiessler (R-F) method, and one for the Monte Carlo method.

This book could not have been completed and published without lots of encouragement and help. First, I sincerely thank Mechanical Department Chairman and Professor Mickael Jackson at the Wentworth Institute of Technology, whose encouragement motivated me to open two technical elective courses about the reliability in mechanical engineering. Second, I sincerely thank Professors Anthony William Duva and Richard L. Roberts for reviewing some of the manuscripts. Third, I sincerely thank Morgan & Claypool Publishers and Executive Editor Paul Petralia for helping with this publication. Finally, I sincerely thank my lovely wife, Suyan Zou. Without her support, I could not have completed this book.

Xiaobin Le
October 2019

CHAPTER 1

Introduction to Reliability in Mechanical Design

1.1 ENGINEERING DESIGN PROCESS

Mechanical engineering is one of the oldest branches of science. Mechanical design is one of the primary purposes of mechanical engineering. It has created a lot of fantastic design projects which greatly benefit human society such as the steam engine, elevator, bridge crane, automobile, train, ship, and airplane. Unfortunately, some failures of mechanical design projects also caused disasters to human society such as the Space Shuttle Challenger Disaster on January 28, 1986, which was mainly due to the failure in mechanical component design and resulted in the death of all seven crew members including five NASA astronauts and two payload specialists. Because of the complexity of modern mechanical design, mechanical design theory has been constantly developed and updated. This book contributes to mechanical design theory by presenting how to conduct reliability-based mechanical design.

The first key topic of mechanical design theory is the description of the engineering design or engineering design process, which is a summary of past success and failure of design experience. This section will briefly describe the definition of engineering design and concisely explain the engineering design process.

There are many definitions of engineering design. The Accreditation Board for Engineering and Technology (ABET) definition of engineering design is: *"Engineering design is the process of devising a system, component, or process to meet desired needs. It is a decision-making process (often iterative), in which the basic science and mathematics and engineering sciences are applied to convert resources optimally to meet a stated objective. Among the fundamental elements of the design process are the establishment of objectives and criteria, synthesis, analysis, construction, testing, and evaluation."* This definition explains the four key aspects of the engineering design.

- Engineering design is a process.

- Engineering design utilizes many different skills with an iterative, decision-making, and systematic approach.

- Engineering design will build an object, including a system, component, or process under some constraints.

- The purpose of the engineering design is to meet the project's required needs.

The most important aspect of engineering design is that engineering design is a process. Why is it a process? Engineering design is a process not because it consists of several steps and might take a long time to complete it, but mainly because the final approval of the design is not by theoretical calculations or numerical simulations, but by the actual testing on the prototype. For example, when we do a complicated math problem, which might need a few hours or a few days, we know the obtained solution is right or not after we finish it because we can plug them in the equations to check the solution. For an engineering design project, even the accurate theoretical calculation and complicated numerical simulation results suggest that the design should be safe and could satisfy the design requirements, the design product cannot be released to mass-production and customers. Engineering design experience has approved that design product without throughout testing on the prototypes might cause significant problems for customers and company.

According to the last few hundred years of engineering successful and unsuccessful experience, lots of different theory about the design process have been proposed for guiding engineers to conduct the engineering design. The five-phase engineering design process proposed by Gerard Voland is one of these good theories. The five-phase engineering design process includes: (1) Phase One: Needs Assessment; (2) Phase Two: Design Specifications; (3) Phase Three: Conceptual Designs; (4) Phase Four: Detailed Design; and (5) Phase Five: Implementation. The detailed information about the five-phase design theory can be found in the book *Engineering by Design*, authored by Gerard Voland [1]. We will provide brief descriptions of the five-phase engineering design process as follows [2].

1.1.1   PHASE ONE: NEEDS ASSESSMENT

The main task in Phase One is to check whether needs are real and feasible or not. The "needs" is a current problem or current unsatisfactory status. The needs can come from personal experience, customers, or society. However, some claiming demands might not be a real need and might fade quickly. Some needs might not be feasible for the project team because it might not be technically feasible or financially viable. For example, many people wish that they could have a device which could directly import the knowledge of books into their brains. This device might be feasible in the future. However, it is certainly not technically feasible now. For another example, constructing a more advanced airplane is always a real need, but it will not be financially viable for a small company because it has limited human resources and funding.

The outcomes of Phase One are as follows. (1) The need is not a real need. No further action is needed. (2) The need is a real need, but it is not technically nor financially feasible for the project team. The need might be stored for future use. No immediate action is needed. (3) The need is a real and feasible need for the project team. Only if the need assessment passes Phase One, will it proceed to Phase Two: Design Specifications. However, the design project or the project team is not still officially established or formed.

1.1.2 PHASE TWO: DESIGN SPECIFICATIONS

The main tasks in Phase Two are: (1) true understanding of the needs; (2) search existing related solutions for the needs; and (3) determine design specifications for the needs. The needs in Phase One, which might come from the sale department, or customers, or society, are generally described by general language. These needs are required to be rephrased by engineering language. Before this, the first step is to have a true understanding of the needs, and then some primary criteria are set up. Before we start to search existing solutions or possible solutions, we must fully and truly understand the need, or the real meanings of the need, or the real requirements from the customers. For example, if customers require us to design transportation for them. If we provide the transportation which is driven by gasoline or diesel engine or electrical engine, we make a big mistake when the power source in that region for the customers is natural gas. The second step is to search existing or possible solutions for the need by using the primary criteria established in the first step. Do not reinvent the wheel. So, the design project for the needs should use up-to-date techniques. After lots of background investigation has been done, and lots of related up-to-date technical information has been collected, the third step is to establish design specifications for the needs. There are several common or general design goals that are usually associated with design projects such as safety, environmental protection, public acceptance, reliability, performance, durability, ease of operation, use of standard parts, minimum cost, minimum maintenance and ease of maintenance, and manufacturability. The design specifications do not mean these common or general goals, which are certainly considered in detail. The design specifications properly established in the third step are: (1) any special statement for the design project and (2) any numerical value related to the design project. Two examples of design specification are: "The device must work properly in a very moist environment" and "the factor of safety is 3.5."

The outcomes of Phase Two are as follows. (1) The needs could be satisfied by an existing product, or profitable product could not be constructed due to the highly competing market. So, no further action is required. The design project will not be set up. (2) Proper design specifications have been established and will be satisfied by design. The design project will be set up, and the project team will be formally formed.

Only after the design specifications are established will the design project and project team be formally assembled. Now the design project will move to Phase Three: Conceptual Design.

1.1.3 PHASE THREE: CONCEPTUAL DESIGN

The main task in Phase Three is to develop a pool of all possible alternative solutions for the design project, and then to generate several best alternative solutions for the design project from the pool of all possible alternative solutions.

Modern design projects are typically complicated with several functions or subsystems. The following four steps can be used to conduct the conceptual design.

Step 1: Decompose a design project into several subsystems or subunits according to their functions. Any design project normally consists of several subsystems/subunits or several different functions. It is very complicated to generate solutions for a unit with several functions. However, it will be relatively easy to generate solutions for a subunit with only one function. So, in this step, the design project will be decomposed into a lot of subsystems/subunits, which have only one key function or performance.

Step 2: Develop a lot of possible options for each subsystem/subunit. A subsystem/subunit with single function or performance can be abstracted into a general catalog, which will be backed up by the whole engineering technique and knowledge. Then, the project team can generate many solutions for each subsystem/subunit. In this step, full information research should be conducted. The good quality of these solutions will determine the quality of the final design solution because the final design solution of the design project will be formed from these solutions.

Step 3: Form a pool of alternatives for the design project. The combination of one option from each subsystem/subunit will form one alternative for the design project. This combination will typically be a large quantity of possible alternative designs for the project. For example, if there are six subsystems, and each subsystem has three solutions, the possible alternatives for a design project will be equal to $3^6 = 729$.

Step 4: Use personal experiences, judgment, or group discussion to choose several best prospective alternatives for the design project. The project team does not have enough time and money to build and test each possible alternative. It is also not necessary. So, the project team will use personal experience, judgment, and group discussion to pick up several best alternatives for further investigations in the next stage.

The outcome of Phase Three: Conceptual Design offers several best alternatives for the design project, from which the final design solution will be selected.

1.1.4   PHASE FOUR: DETAILED DESIGN

The main tasks in Phase Four are: (1) create a virtual product and run the virtual experiments on the virtual product; (2) choose the final design option for the design project; and (3) test and modify the prototype and finalize the final design option for the design project.

In modern engineering society, lots of modern advanced tools are available for engineering design. For example, computer-aided design (CAD) software can be used to build a virtual component (VC), virtual assembly (VA), and virtual product (VP). A VC has the same dimensions, the same material, and the same behaviors of a real component. So, it looks like the real component and behaves like the real component but is stored in a digital form. A VA is a digital form of a real assembly. The VP is a digital form of a real product. Lots of engineering simulation software such as Solidworks Simulation and ANSYS, which are commercial finite element analysis software, can be used to test the functions and stress/strain of a virtual component, or assembly, or product under different loading conditions. A lot of different computer-aided manufacturing (CAM) can be used to animate the manufacturing process of a VC to check the feasibility of manufacturing.

The first task in Phase Four is to construct VCs, VAs, and VPs, and then use available computer software to conduct the numerical simulation to check the performances and functions of different design options. Only some key components or key sub-assemblies might be manufactured and experimented on the prototypes of components or sub-assemblies for checking the performance of functions. The contents of this book will be an engineering design theory to determine the dimension of a component with required reliability, that is, VC.

The second main task is to determine the best design option after performance and functions of several design options under consideration have been evaluated through virtual tests, that is, numerical simulation on VCs, VAs, and VPs. Since design project has several design specifications and functions, systematic evaluation methods such as Decision Matrix, which is fully explained in the book *Engineering by Design* [1], could be implemented to choose the best option.

The last main task is to build the prototype of the final design option and then thoroughly test and modify it until all required design specifications, performance, and functions are properly satisfied by the final design option.

The outcome of Phase Four is a tested and approved final design option, which is ready to be released for production.

1.1.5   PHASE FIVE: IMPLEMENTATION

The last phase in the engineering design process is the implementation, which mainly means transforming a design into reality. The main task in Phase Five is to prepare a complete set of engineering documentation, such as drawing, part list, operation and maintenance manual, quality control procedures, manufacturing procedures, and manufacturing tool and fixture design. The complete set of engineering documentation is the technical information by which the final design option can be manufactured or duplicated. The outcome of Phase Five is a complete set of engineering documentation.

An iterative-interactive five-phase engineering design process shown in Figure 1.1 includes six items. The center is the project team. The project team is the core of any engineering design process. The iterative-interactive process and all other design activities are carried out by and through the project team. The other five elements are needs assessment (Phase One), design specification (Phase Two), conceptual design (Phase Three), detailed design (Phase Four), and implementation (Phase Five). An engineering design project has its own life—it starts at Phase One: Needs Assessment and ends at Phase Five: Implementation. The flow chart of the engineering design process is naturally from Phase One, to Phase Two, to Phase Three, then to Phase Four, and finally to Phase Five. During this design process, interactive activities could

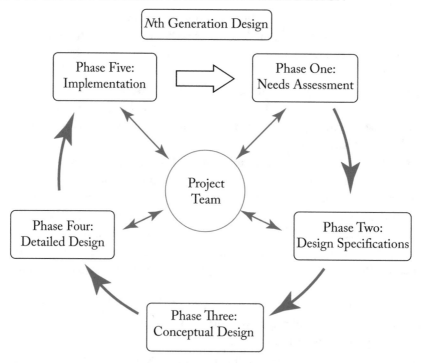

Figure 1.1: The iterative-interactive five-phase design process.

happen between any phase. For example, during Phase Four the project team could modify final design option based on virtual numerical simulation and the physical testing on the prototype. For another example, during Phase Three, the project team could back to work on Phase Two to make some modifications. Description and procedure of the engineering design process such as the five-phase engineering design process is the summary of past successful and unsuccessful design experience. It is a piece of important and critical knowledge and skill. The project team should follow it. It does not mean that following the five-phase design process will guarantee to have a successful design result for a design project. It only suggests that there will be a very high possibility that design project will end with a failure or high cost or long period of time if the procedure of engineering design process does not follow.

1.2   FAILURES IN ENGINEERING DESIGN

Mechanical components are designed to execute required performances/functions under the specified working environment and loading conditions (design specifications) within the specified life of service. However, due to the practical and economic limitations, a perfect design does not exist. So, some of the mechanical components will certainly fail. Failure is defined as a phenomenon that mechanical components cannot satisfy the design specifications or not provide the required performance. For example, if a bar under the rated loading fractures, the bar is said to be a failure. If a shaft under the specified working environment and loading condition has excessive deflection, which might affect the proper gear engagement on the shaft, the design of the shaft is said a failure. For another example, if a camera can take a photo, but the image of the photo is not clear, the camera is said to be a failure.

The well-known failure-rate curve [3], known as the bathtub curve, is widely utilized to describe and explain the failure of electronic, mechanical, and electro-mechanic components. The schematic of a typical bathtub cure of the failure rate vs. time is shown in Figure 1.2. The horizontal axis is the time of the component in service. The vertical axis is the failure rate, which is defined as the frequency failures per unit of time. For example, if a design component with 20,000 units in service for 5,000 hr have 254 failures, the failure rate of this design component will be: (254/20000)/(5000), that is, 2.54×10^{-6} *failure per hour* or 2.54 *failure per million hour*. The typical bathtub curve consists of three different stages. The first stage is the early stage of the product life known as the infant mortality stage, where there is a rapidly decreasing failure rate. The failures in the first stage are mainly due to manufacturing defects and poor-quality control procedures. These failures can be prevented and eliminated if the careful manufacturing and proper quality controls are applied during the production. As these defective components are replaced/repaired, the failure rate decreases as time progresses in the first stage. The second stage has an almost constant failure rate, which is known as the useful life stage. The constant failure rate of the product indicates that there is no dominant failure mechanism to induce a failure; that is, the failure is mainly due to random causes. For example, a mechanical component failed due to accidental overload when the material strength of this component was in the lower end of such material's normal strength range. Products' life should be designed to be in the second stage. The failure rate of some mechanical products in the second stage are listed in Table 1.1 [3, 4]. The failure rates listed in the table represent the current industrial product design level with both practical and economic considerations. The third stage is known as the wear-out stage, where there is a rapidly increasing failure rate. The dominant failure mechanism of the products in this stage is "wear-out," such as the cumulative irreversible fatigue damage due to continuous cyclic

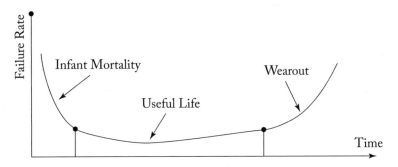

Figure 1.2: The bathtub curve.

Table 1.1: Failure rate of some mechanical products [3, 4]

Mechanical Component	Failures Per Million Hours	Mechanical Component	Failures Per Million Hours
Accelerometer	35.1	Gear	0.17
Actuator	50.5	Gear shaft	6.7
Air compressor	6.0	Gyroscope	513.9
Air pressure gauge	2.6	Heat exchanger	1.1
Ball-bearing	1.1	Hydraulic valve	9.3
Boiler feed pump	0.42	O-ring	2.4
Brake	4.3	Roller bearing	0.28
Clutch	0.6	Shock absorber	0.81
Differential	15.0	Spring	5.0
Fan	2.8	Storage tank	1.6
Gasket and seal	1.3	thermostat	17.4

loads. It is strongly recommended that the designed products' life would not be extended into the wear-out stage because the failure rate would be very high.

Mechanical components have many different failure modes such as static failure, fatigue failure, creep failure, corrosion failure, wear failure, instability failure, and excessive deflection failure [5–7]. However, static failure, and excessive deflection failure, and fatigue failure will be focused on in this book.

- The static failure. When a component's maximum stress due to working load exceeds material strengths such as yield strength and ultimate strength, the component is defined as a static failure. For example, a component of brittle material will fracture when the component's maximum stress exceeds the material's ultimate strength. For another example, a component of ductile material will have excessive deflection and lose the capability of carrying out working load when the component's stress exceeds the material's yield strength.

- Excessive deflection failure. Mechanical systems typically have at least one moving component. When the excessive deflection of a component causes the mechanical system to fail to satisfy the required performance and design specifications, this is defined as an excessive deflection failure. For example, excessive deflection of a shaft might cause big noise during the gear engagement, which might exceed the permissible sound level.

- The fatigue failure. When a component is subjected to cyclic load, the component will be gradually degraded due to fatigue damage. The fatigue damage is irreversible and is accumulated during the service. After the accumulated fatigue damage reaches a critical value, the component will fail by fracture, even the magnitude of the nominal maximum cyclic stress is far below the material's yield strength. This failure is defined as fatigue failure. In reality, all mechanical components are subjected to cyclic load due to the continuously repeated performance or moving components, or mechanical vibrations. So, it is well known that more than 90% of mechanical metal components fail due to fatigue failure. For example, ball bearings or spur gears will undoubtedly fail when the service time is big enough.

1.3   UNCERTAINTY IN ENGINEERING

Uncertainty is the lack of certainty, a situation that is impossible to describe and predict exactly by one outcome or one numerical value. Uncertainty exists in most phenomena and events observed worldwide. In practice, the various forms of uncertainty can be classified into two categories: physical uncertainty and cognitive uncertainty [3, 8, 9].

- Physical uncertainty results from the fact that a system can behave in random ways and is associated with the state of nature. This type of uncertainty is the inherent randomness that exists in all physical parameters, so it is an irreducible uncertainty. Physical uncertainty will be the focus of this book. Uncertainties of the design parameters in the mechanical design are physical uncertainty. For example, material mechanical properties of the same brand material such as yield strength, ultimate strength, and Young's modulus will inevitably vary due to small variations in chemical composition, the temperature in heat treatment, and non-homogeneous temperature field during solidification. So, the material mechanical properties are physical uncertainty. The geometric dimensions of components are physical uncertainty because no two components can be made identical due to tool wear, errors in measurement, machine tool vibrations, or the resistance of the shaft materials to cutting. For another example, external load or operation patterns in the mechanical design are physical uncertainty because the actual external load of the mechanical system in real service are unpredicted and varies from one working condition to another condition, from one service to another. The approach in mechanical design to deal with these physical uncertainties will be reliability-based mechanical design, which is the main topic of this book and will be discussed and explored in detail later.

- Cognitive uncertainty results from the lack of knowledge about a system and is associated with our interpretation of the physical world. Cognitive uncertainty describes the inherent vagueness of the system and its parameters. For example, the outcome of the quality of machined mechanical components by machine operators whose skill and experience in machining are unknown is cognitive uncertainty. When machine operators' skills and experience are fully known, the outcome of the quality of machined mechanical components by those operators will be a physical uncertainty.

1.4   DEFINITION OF RELIABILITY

When uncertainties of main design variables are taken into consideration during mechanical design, reliability will be a relative measure of the performance of a product. Although there is a consensus that reliability is an important attribute of a product, there is no universally accepted definition of reliability. This book will use the following definition of reliability.

Reliability,   denoted by R, is defined as the probability of a component, a device, or a system performing its intended functions without failure over a specified service life and under specified operation environments and loading conditions.

Phase Two of the engineering design process, as discussed in Section 1.1, is to determine and specify the design specifications of the product. The design specifications include the intended functions, service life, operational environments, loading conditions, and the required reliability. There are lots of different possible failure, as discussed in Section 1.2. This book will only focus on three typical mechanical components' failures: static failure, excessive deflection failure, and fatigue failure. Therefore, the reliability of a component can be expressed by the following equation:

$$R = P(S \geq Q), \tag{1.1}$$

where S is a component material strength index, which could be material strength such as yield strength, ultimate strength, fatigue strength, or allowable deflection. Q is a component loading index, which could be maximum stress, accumulated fatigue damage, or maximum deflection of a component.

$P(S \geq Q)$ means the probability of the status that component can perform its intended functions without failure. R is the reliability and is equal to this probability. The physical meaning of reliability is the percentage of components working properly out of the total of the same components in service. For example, if 10,000 of the mechanical shafts with the designed reliability 0.99 for the service life of one year are in service, the reliability 0.99 of these shafts indicate that $0.99 \times 10000 = 9900$ of these mechanical shafts are expected to work properly at the end of one-year service. One hundred of these mechanical shafts might fail at the end of one-year service.

Reliability is an important attribute of a component and a relative measure of the component status through the comparison of materials strength index with component loading index within the service life. In other words, the reliability of a component is a function of materials properties, loading conditions, component geometric dimensions, and service life. Since reliability R is expressed by probability, reliability is not an attribute of a specific component, but an attribute of the batch of same designed components. For example, a company has designed and sold 10,000 unit of the designed component with a reliability 0.99. Supposed we purchased one of the components. We cannot claim that the component has a reliability of 0.99 because the reliability 0.99 is the attribute of the batch of 10,000 units of the same designed components.

After the definition of reliability is defined, it is easy to define the probability of failure of a component.

Probability of failure, denoted by F, is defined as the probability of a component, a device, or a system failing to perform its intended functions over a specified service life and under specified operation environments and loading conditions.

The probability of failure, F, can be expressed by the following equation:

$$F = P(S < Q). \tag{1.2}$$

The sum of the reliability and the probability of failure should be equal to 1, that is,

$$F + R = 1. \tag{1.3}$$

Example 1.1

It is assumed that a company has designed and sold 10,000 units of the designed component which has a reliability of 0.99 with a service life of 5 years. The record of the service information of these 10,000 units during 5 years in the service is shown in Table 1.2. Estimate the reliability of the components at the end of each service year and actual reliability of the components with the service life of five years.

Table 1.2: The record of the service information of the designed components

Service Years	First Year	Second Year	Third Year	Fourth Year	Fifth Year
Number of Failures	11	15	13	17	27

Solution:

From Table 1.2, at the end of the first-year service, 11 of the components fails, and $10,000 - 11 = 9,989$ of components still work properly. According to the definition of the reliability and the probability of failure, we can estimate the reliability R_1 and the probability of failure F_1 of the components when they are in service of one year:

$$R_1 = \frac{9,989}{10,000} = 0.9989, \quad F_1 = \frac{11}{10,000} = 0.0011.$$

The reliability and the probability of failure of the components at the end of each service year can be calculated accordingly and is listed in Table 1.3.

From Table 1.3, the actual reliability of the components with a service life of 5 years is 0.9917, which is larger than the required reliability 0.99. This result indicates that the components have been designed properly. ∎

Table 1.3: Estimation of the reliability and the probability of failure of the components

Service Years	One Year	Two Years	Three Years	Four Years	Five Years
Number of Working Components	9989	9974	9961	9944	9917
Reliability	0.9989	0.9974	0.9961	0.9944	0.9917
Cumulative Number of Failures	11	26	39	56	83
Probability of Failure	0.0011	0.0026	0.0039	0.0056	0.0083

Example 1.2

It is assumed that a company designs and manufacture a mechanical unit with a reliability 0.95 for a service life of 2 years. Estimate how many of the mechanical units will fail at the end of two years if 50,000 of the mechanical unit are in service.

Solution:

The reliability of the mechanical unit is 0.95. So, the probability of failure of the mechanical unit will be:

$$F = 1 - R = 1 - 0.95 = 0.05.$$

According to the definition of the probability of failure, the estimation of the numbers of failures of the mechanical units at the end of 2 years in service will be:

$$n = F \times 50,000 = 0.005 \times 50,000 = 2,500.$$

∎

1.5   IMPORTANCE OF RELIABILITY

A component, device, or system is designed per specifications to properly and safely perform its intended functions. However, throughout the history of engineering design some of them failed to perform the intended functions. For instance, some caused disasters such as the Space Shuttle *Challenger Disaster*, which happened on January 28, 1986, causing a financial loss of around $3.2 billion as well as the tragic human loss of seven astronauts. The factor of safety, is defined as the ratio of component material strength index, such ultimate strength, to the component actual stress index such as component maximum stress. However, the intention of the factor of safety is never used to predict or estimate the likelihood of component failure. The reliability-based mechanical component can address this issue. It is commonly agreed that reliability in engineering design was initiated and established by the AGREE report [10] in 1957. The method of reliability in engineering has become extremely important. The importance of reliability in engineering design can be expressed and explained through the following four aspects.

- Reliability in engineering design is a more scientific and advanced relative measure of the status of a component, device, or system. In reality, all design parameters such as material strength, working load, and component geometric dimensions have some inherent uncertainty. Per the definition of reliability, reliability considers all these uncertainties of design parameters in engineering design and systematically links them together through a probability theory to conduct engineering design.

- Reliability in engineering design is a more practical measure of the status of a component, device, or system because it admits that any designed component could fail no matter how we design it. The reliability R is designed into the components per design specification, is an attribute of the component, and can be used to predict the percentage of failure of the designed component. The percentage of failure of the component will be equal to $(1 - R)$.

- Reliability is directly and tightly related to the reputation of a company and a product. In today's competitive market, every device and system is expected to perform satisfactorily throughout its expected life span. Usually, the value of reliability is proportional to the price of the product. The higher reliability the product has, the higher the cost needed to manufacture it. Product with higher reliability might affect company markets. However, the lower the reliability of the product indicates a higher percentage of failure. Although the manufacturer gives a warranty to cover the failures of the product during its early stages of life, too many failures during the warranty period not only cause inconvenience to the customer and high cost of repair to the manufacturer but also mean a loss of reputation and market share.

- Reliability in engineering design is a systematic tool for designing a complicated product or system with the required reliability. Generally, no device or system will perform reliably unless it is designed specifically for reliability. The complicated system typically includes many subsystems, and each subsystem can have thousands of components. Reliability in engineering design intends to design the reliability into each component or to predict the percentage failure of each component. The proper arrangement or allocation of reliability into each component can design each subsystem with the required reliability.

1.6 RELIABILITY HISTORY

Reliability in engineering design is an application of probability theory. Ancient mathematicians such as Italian mathematician Gerolamo Cardano in the 16th century, and French mathematicians Pierre de Fermat and Blaise Pascal in the 17th century, studied and researched probability theory, which was mainly about gambling. However, the reliability in engineering emerged as a separate discipline in the 1950s due to the extensive research and study on military electronic equipment.

During World War II, Germans applied basic reliability concepts to improve the reliability of their V1 and V2 rockets [3]. The following are some unbelievable facts about electronic equipment during World War II [3].

- During 1941–1945, 60–75% of vacuum tubes in communication equipment failed.

- During 1941–1945, nearly 60% of the airborne equipment shipped to the Far East was damaged on arrival.

- During 1941–1945, nearly 50% of the spare parts and equipment in storage became un-serviceable before they were ever used.

- In 1947, nearly 70% of the electronic equipment possessed by the Navy was not operating properly.

Such extremely high percentages of electronic equipment failures, in reality, resulted in great attention and extensive research activities in the United States. In 1950, the Air Force formed the ad hoc Group on Reliability of Electronic Equipment to study the situation and recommend measures that could increase the reliability of equipment. In 1951, the Navy began an extensive and lengthy study on vacuum tubes. In 1951, the Army started a similar investigation. Due to such an extensive study, in 1952, the Department of Defense established the Advisory Group on Reliability of Electronic Equipment (AGREE) to coordinate the research activities of the Air Force, Navy, and Army. In June 1957, AGREE published its first report: "Reliability of Military Electronic Equipment" [10]. Two of their many conclusions and recommendations in the AGREE report were: (1) reliability testing must be made an integral part in the development of a new system and (2) procuring agencies should accept the equipment only after getting the reliability demonstrated by a manufacturer. It is widely accepted that this AGREE first report in 1957 is the foundation of the reliability in engineering design.

The extensive research on the reliability of electronic equipment in the 1950s formed the first set of reliability-related standards. Some examples of these military standards are listed in the following.

- In 1952, the AGREE established a military standard MIL-STD-781 "Reliability Qualification and Production Approval Test," which was revised as MIL-STD-781b in 1967.

- In 1955, the AGREE established military standards MIL-STD-441 "Reliability of Military Electronic Equipment."

- In 1961, MIL-STD-756 for reporting prediction of weapons' system reliability.

- In 1965, the DOD (Department of Defense) military standard MIL-STD-785 "Reliability Program for Systems and Equipment," which was revised in 2008.

Reliability in mechanical engineering design is founded based on similar concepts and principles which were established through the extensive research on military electronic equipment. The following are several notable events for reliability in mechanical engineering design.

- In 1951, Weibull, W. of the Swedish Royal Institute of Technology published a statistical distribution for material strength [11]. This distribution is called Weibull distribution and has played an important role in the development of reliability in mechanical engineering.

- In 1968, Professor Edward B. Haugen published the book *Probability Approach to Design* [12], which was directedly focused on mechanical design with reliability.

- In 1972, Professor A. D. S. Carter published a book *Mechanical Reliability* [13].

- In 1977, Dr. D. Kececioglu published a paper "Probabilistic design methods for reliability and their data and research requirements" to present the approach for dealing with reliability in fatigue failure [14].

Nowadays, reliability in engineering design has become an important concept and tool for mechanical engineering design.

1.7   RELIABILITY VS. FACTOR OF SAFETY

The differences between reliability and factor of safety will be very clear after Chapter 4 has been read. Before that, we will use some simple examples to explain their differences.

In the traditional mechanical design approach, the factor of safety is typically defined as the ratio of component' strength to the maximum component stress induced by the operational load. For example, if the failure mode of the component is a static failure, the factor of safety is:

$$n = \frac{\overline{S}}{\overline{Q}} > 1, \tag{1.4}$$

where $\overline{S}$ is the average of the component's strength index such as tensile yield strength or ultimate tensile strength for static failure design. $\overline{Q}$ is the average of the component's maximum stress, which can be determined by the component's geometrical dimensions and operational loading. n is the factor of safety, which links together component's material strength, component geometrical dimensions, and operational loading for component design.

Reliability of a component is defined as the probability of a component, a device or a system performing its intended functions without failure over a specified service life and under specified operation environments and loading conditions. The mathematical equation for the reliability is shown in Equation (1.1) and is repeated here:

$$R = P(S \geq Q), \tag{1.1}$$

where S is the component's strength index such as tensile yield strength or ultimate tensile strength for static failure design. Q is the component's maximum stress, which can be determined by the component's geometrical dimensions and operational loading. R is reliability.

Both the reliability and the factor of safety serve the same purpose and are a measure for creating the design equations. However, the key differences between them are:

- the factor of safety is a deterministic approach in which all design parameters are treated as deterministic values. The factor of safety intends partially to consider the uncertainty of the design parameters; and

- the reliability is a probabilistic approach in which all design parameters are treated as random values. The uncertainties of the design parameters are assessed by reliability.

The following example can explain in detail the similarities and differences between the reliability and the factor of safety.

Example 1.3

In the traditional design approach with a factor of safety, the design parameters for three design cases with different materials and different operation loading are treated as deterministic values and are listed in Table 1.4. When the uncertainties of the strengths and stresses of the three design cases are considered, the strengths and stresses of the components of the same design cases are treated as normal distribution random variables (note: the normal distribution will be discussed in Section 2.12.4). The corresponding distribution parameters are also listed in Table 1.2:

1. conduct the design check, that is, calculate the factor of safety and the reliability of components; and

2. discuss the results.

Table 1.4: The design parameters of three design cases

| Case # | Traditional Approach with a Factor of Safety | | The Probabilistic Approach with a Reliability | | | |
| | | | Strength S (normal distribution) | | Stress Q (normal distribution) | |
	Strength	Stress	Mean μ_S	Standard Deviation σ_S	Mean μ_Q	Standard Deviation σ_Q
1	100 (ksi)	80 (ksi)	100 (ksi)	5 (ksi)	80 (ksi)	5 (ksi)
2	100 (ksi)	80 (ksi)	100 (ksi)	5 (ksi)	80 (ksi)	20 (ksi)
3	100 (ksi)	50 (ksi)	100 (ksi)	30 (ksi)	50 (ksi)	30 (ksi)

Solution:

1. The factor of safety and the reliability.

   The factor of safety of three design cases can be calculated per Equation (1.3) and are listed in Table 1.5. The reliability can be calculated according to Equation (1.1), which will be discussed in detail in Section 2.12.4. The random event $(S \geq Q)$ is the same random event $(S - Q \geq 0)$. Let us use Z to represent the new random variable $S - Q$, that is, $Z = S - Q$. Equation (1.1) can be rearranged as:

   $$R = P(S \geq Q) = P(S - Q \geq 0) = P(Z \geq 0).$$

   Since both strength S and stress Q are normal distributions, the random variable Z will also be a normal distribution. The mean μ_Z and standard deviation σ_Z of Z can be calculated by the means and the standard deviations of S and Q. They are:

   $$\mu_Z = \mu_S - \mu_Q; \qquad \sigma_Z = \sqrt{(\sigma_S)^2 + (\sigma_Q)^2}.$$

   After the mean and standard deviation of the normally distributed random variable Z are determined, the reliability can be directly calculated based on $R = P(Z \geq 0)$ and are listed in Table 1.5. (Note: The calculation procedure will be discussed in detail in Section 2.12.4.)

Table 1.5: The factor of safety and the reliability of three design cases

Case #	The Factor of Safety Approach			The Reliability Approach		
	Strength	Stress	Factor of Safety	Mean μ_Z	Standard Deviation σ_Z	Reliability
1	100 (ksi)	80 (ksi)	1.25	20 (ksi)	7.1 (ksi)	0.9977
2	100 (ksi)	80 (ksi)	1.25	20 (ksi)	20.6 (ksi)	0.8340
3	100 (ksi)	50 (ksi)	2	50 (ksi)	42.4 (ksi)	0.8807

2. Discuss the results.

   - Both the factor of safety and the reliability are the measure of the status of safety of the components. However, the reliability R not only predicts the status of safety of the component but also indicates failure probability F, which is equal to $1 - R$. For example, from Table 1.5, the reliability of components in Case #1 is 0.9977, and the failure probability of a component is $1 - 0.9977 = 0.0023 = 0.23\%$. The factor of safety cannot provide any information about the possible failure of a component.

- From Table 1.5, both Cases #1 and #2 have the same factor of safety: 1.25. The same value of the factor of safety implies that both Case #1 and Case #2 will have the same measure of the status of safety. However, when the reliability approach is used to check the status of safety of Case #1 and Case #2, the reliabilities of these two designs are quite different, as shown in Table 1.5. Case #1 has a reliability 0.9977, and Case #2 has a reliability 0.8340 only. The cause for this inconsistent result is due to the uncertainty of design parameters. The traditional design approach with the factor of safety cannot quantitively consider the effect of uncertainty. The design approach with reliability does consider the effects of uncertainty.

- From Table 1.5, the factor of safety in the design Case #3 is 2 and is larger than the factor of safety 1.25 of the design Case #1 from the table. According to the meaning of factor of safety, this indicates that the components from the design Case #3 should be relatively safer than the components from the design Case #1. However, the reliability of the components for the design Case #3 is much less than the reliability of the component for the design Case #1 from Table 1.5. The cause for these contradictory conclusions is mainly due to the uncertainty of design parameters. So, a higher factor of safety does not guarantee a much safer component. However, higher reliability will certainly guarantee just that.

- From Table 1.4, the simple information about the design parameters, that is, deterministic values, are required when the design approach with a factor of safety is used for component design. However, when the design approach with reliability is used for component design, a large amount of information about design parameters are needed because the type of distributions and corresponding distribution parameters are required. ∎

In summary, both the factor of safety and the reliability are the measure of the status of safety of a component. Both are successfully used for mechanical component design. The advantages of the factor of safety are simple and do not require much information about design parameters. The disadvantages are: (1) it cannot be used to explain possible component failure; (2) the higher the factor of safety of components does not guarantee that it will be much safer; and (3) it cannot include the effects of uncertainty of the design parameters. The advantages of reliability are: (1) it not only indicates the probability of safe components, but also indicates the probability of component failure; (2) the higher reliability of a component certainly guarantees that it is much safer; and (3) the approach with reliability can fully consider the effects of uncertainties of design parameters. The main disadvantage of the design approach with reliability is that much more information or a large amount of data about uncertainties of design parameters are required. Without reliable descriptions of uncertainties of design parameters, the implementation of the design approach with reliability is meaningless.

1.8   SUMMARY

Engineering design is a process in which a design idea is checked for its feasibility, rephrased by engineering technical language, designed, and then manufactured into a product that will serve the society. The engineering design is a process not only because it takes a period of times to be completed, but also because the prototype of design must be built and fully tested before it can be released to the society. One description of the engineering design process is the five-phase engineering design process—Phase One: Needs Assessment; Phase Two: Design Specifications; Phase Three: Conceptual Designs, Phase Four: Detailed Design; and Phase Five: Implementation. The contents of this book are the key techniques required to design components in Phase Four: Detailed Design.

Reliability R is defined as the probability of a component, a device or a system performing its intended functions without failure over a specified service life and under specified operation environments and loading conditions. The reliability-based mechanical design is the main topic of this book. It uses the reliability to replace the factor of safety to form a design governing equation for linking all design parameters together. One dilemma for the traditional design theory with the factor of safety in mechanical component design is that it cannot provide any tool to assess the component failure, which is a reality in industries. The reliability-based mechanical design is an advanced design theory and can provide a tool to solve this dilemma. A few highlights about the reliability-based mechanical design are follows.

- All main design parameters such as materials strengths, component dimensions, external loadings, and the loading-induced component stresses are treated as random variables for considering their uncertainties.

- The component is designed with required reliability. Its physical meaning is the percentage of components working properly under specified operation environments and loading conditions over a specified service life.

- Probability of failure F of the component will be $(1 - R)$ and indicates the percentage of components which cannot work properly under specified operation environments and loading conditions over a specified service life.

- The higher the reliability a component has, the safer it will be. However, the higher reliability of a component also implies a higher cost of the component.

1.9   REFERENCES

[1] Voland, G., *Engineering by Design*, Pearson Prentice Hall, 2004. 2, 5

[2] Le, Xiaobin, Anthony, D., Roberts, R., and Moazed, A., Instructional methodology for capstone senior mechanical design, *ASEE International Conference*, Vancouver, BC, Canada, June 26–29, 2011. 2

[3] Rao, S. S., *Reliability Engineering*, Pearson, 2015. 7, 8, 9, 14

[4] Reliability analysis center, non-electronic parts reliability data, NPRD-1, *Reliability Analysis Center*, Rome Development Center, Griffiss Air Force Base, New York, 1978. 7, 8

[5] Cater, A. D. S., *Mechanical Reliability and Design*, John Wiley & Sons, Inc., New York, 1997. 8

[6] Modarres, M., Kaminskiy, M., and Krivtson, V., *Reliability Engineering and Risk Analysis: A Practical Guide*, 2nd ed., CRC Press, 2010. DOI: 10.1201/9781420008944.

[7] Dasgupta, A. and Pecht, M., Materials failure mechanisms and damage models, *IEEE Transactions on Reliability*, 40(5), 531–536, 1991. DOI: 10.1109/24.106769. 8

[8] Choi, S.-K., Grandhi, R. V., and Canfield, R. A., Reliability-based structural design. DOI: 10.1007/978-1-84628-445-8. 9

[9] Vinogradov, O., *Introduction to Mechanical Reliability, a Designer's Approach*, Hemisphere Publishing Corporation, 1991. 9

[10] AGREE Report, Advisory Group on Reliability of Electronic Equipment (AGREE), Reliability of Military Electronic Equipment, Office of the Assistant Secretary of Defense (Research and Engineering), Department of Defense, Washington, DC, 1957. 13, 14

[11] Weibull, W., A statistical distribution function of wide applicability, *Journal of Application Mechanics Transactions on ASME*, 18(3): 293–297, 1951. 15

[12] Haugen, E. B., *Probability Approach to Design*, New York, Wiley, 1968. 15

[13] Carter, A. D. S., *Mechanical Reliability*, New York, Wiley, 1972. DOI: 10.1007/978-1-349-18478-1. 15

[14] Kececioglu, D., Probabilistic design methods for reliability and their data and research requirements, *Failure Prevention and Reliability, ASME*, pp. 285–305, 1977. 15

1.10   EXERCISES

1.1.   List at least two different definitions of engineering design and discuss their similarities and differences.

1.2.   Use your statement to explain what engineering design is.

1.3.   What is the five-phase engineering design? What are the main tasks of each phase?

1.4. Use one of your design projects to check whether you used all five phases. If not, why or what happened?

1.5. Why is the engineering design a process?

1.6. List at least two possible needs for engineering design and then run the needs assessment.

1.7. Use one of your design examples to describe how you conduct a conceptual design.

1.8. List four examples of failures that you know. What are the types of failures? What causes failures?

1.9. Describe and explain your understanding of a typical bathtub curve of the failure rate.

1.10. What is a static strength failure?

1.11. What is fatigue failure? Explain the differences between static failure and fatigue failure.

1.12. List one example of excessive deflection failure and provide some possible solutions to deal with the issue.

1.13. What is uncertainty in engineering? List and explain at least three examples of typical design parameters.

1.14. What causes the uncertainty of the ultimate tensile strength of a material?

1.15. Explain the uncertainty related to geometrical dimensions of components.

1.16. Use one example to describe and explain uncertainty related to component failures.

1.17. What is the factor of safety? What is the physical meaning of the factor of safety? Can you use the factor of safety to predict the component failure? Why?

1.18. List and explain the definition of reliability in engineering design.

1.19. What are the similarities and differences between the factor of safety and reliability?

1.20. What is the reliability-based mechanical design? What are the key differences between the traditional mechanical design theory and reliability-based mechanical design?

1.21. Does a component with a higher factor of safety imply that the component will have higher reliability? If not, can you explain it?

1.22. Describe and explain the importance of reliability in engineering design.

1.23. Briefly describe the reliability history. What do you learn from this history?

C H A P T E R 2

Fundamental Reliability Mathematics

2.1 INTRODUCTION

This chapter will first briefly discuss fundamental concepts of set theory and probability, then the probability density function (PDF) and the cumulative distribution function (CDF) of random variables, and finally typical types of distributions and the goodness-of-fit test. The focus of these discussions is not to explain these in a mathematical approach, but how to implement the probabilistic concepts in the reliability-based mechanical design for design engineers. The calculation of the PDF and the CDF will be mainly through Excel or MATLAB®.

2.2 EXPERIMENT, OUTCOME, SAMPLE SPACE, AND EVENT

Every physical parameter in engineering design or real life has a certain element of uncertainty. For example, a wind speed against a building, the weight of a lifted object of a bridge-crane, and the diameter of a machined shaft cannot be predicted exactly. The terms of experiment, outcome, sample space, and the event can be used to describe these phenomena.

Experiment refers to an act of doing something, but the results of it cannot be predicted exactly before the action has been completed. For example, conducting a material tension test is an experiment. Before the tensile test is completed, the actual yield strength of a test specimen is not known. Conduction of 2 plus 2 is not an experiment, because the result of it can be predicted. For another example, tossing a coin or rolling dice is an experiment.

Sample point or outcome refers to the single result of an experiment. For example, rolling a die will have six possible outcomes, which are 1, 2, 3, 4, 5, and 6. For tossing a coin, the occurrence of "head" is a sample point or an outcome.

Per the definition of an experiment, an experiment must have at least two outcomes but also can have an infinite number of possible outcomes. For example, tossing a coin will only have two outcomes: head or tail. The diameter of a qualified shaft with a dimension $\varnothing 1.000 \pm 0.005''$ will have infinite possible outcomes, in which the value of the diameter can be any numerical value between $0.995''$ and $1.005''$.

An event refers to a single outcome or a group of outcomes of an experiment. For example, the even number of rolling a die, that is, {2,4,6} is an event, which includes three possible outcomes. The number 4 of rolling die is also an event, which has only one outcome. For another example, that the yield strength is more than 32 ksi for a material tensile test is an event, which includes infinite possible outcomes.

Sample space is defined as the event that includes all the possible outcomes of an experiment. For example, the sample space of rolling a die consists of numbers 1, 2, 3, 4, 5, and 6. The sample space of tossing a coin will consist of "head" and "tail." The sample space of a material tensile test experiment for yield strength will consist of the infinite possible outcomes, that is, {yield strength $\geq$ 0}.

Mutually exclusive events are the events that cannot happen at the same time. For example, the event of the tail and the event of the head in tossing a coin are mutually exclusive. The event of a failed component and the event of a safe component are mutually exclusive events too because a component cannot be a safe status or a failure status at the same time. For another example, the event of a number less than 3 and the event of even number in rolling die are not mutually exclusive events. If number 2 in an experiment of rolling a die happens, both events happen.

2.3   SET THEORY

The set theory can conveniently describe the operation on the outcomes or events of an experiment and will help us smoothly to understand the concepts and related simple operations of probability.

A set, denoted by a capital letter such as A, is a well-defined collection of objects so that for any given object we can say whether or not it belongs to the set. A set is the equivalent term of the event. In this book, the object means sample points or outcomes of an experiment. A set can be expressed by braces containing the specified sample point or simple points. For example, in rolling a die experiment, a set containing numbers 1 and 4 can be expressed as $A = \{1, 4\}$, where the capital symbol A is the name of this set and $\{1, 4\}$ represents the collection of sample point 1 and sample point 4. For an experiment of ultimate tensile strength S_u, $B = \{5 \text{ ksi} < S_u < 20 \text{ ksi}\}$ is a set. The set B includes the sample point of the ultimate tensile strength is larger than 5 ksi and less than 20 ksi.

A universal set, denoted by Greek letter Ω, is a collection of all possible sample points of the experiment. For example, the universal set of rolling dice is $\Omega = \{1, 2, 3, 4, 5, 6\}$. For another example, the universal set of the status of a component is $\Omega = \{safe, failure\}$.

An empty set or **null set**, denoted by the symbol $\emptyset$, is a set containing no sample point of the experiment. A null set can be expressed by $\emptyset = \{\ \}$. The creation of a null set is mainly for the set operations.

The purpose of introducing some basic operations of the set is to help us to understand some basic calculations of probability. We will discuss the following basic operations of the set.

Union of two sets. Union of two sets A and B, denoted $A \cup B$, is the set of all objects that belong to A, or B or both. The union operation can be graphically represented by the Venn diagram in Figure 2.1.

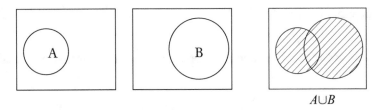

Figure 2.1: Union of two sets A and B.

For example, in rolling a dice experiment, if the set A and B are: $A = \{1, 2\}$, and $B = \{5\}$, the union of two sets A and B will be $A \cup B = \{1, 2, 5\}$.

The intersection of two sets. The intersection of two sets A and B, denoted as $A \cap B$, is the set of all objects that belong to both set A and set B. The intersection of two sets can be graphically represented by the Venn diagram in Figure 2.2.

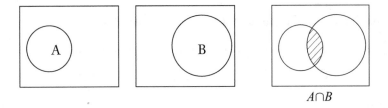

Figure 2.2: Intersection of two sets A and B.

For example, the intersection of $\{1, 2, 3\}$ and $\{3, 4, 6\}$ is the set $\{3\}$. The intersection of $\{1, 2, 3\}$ and $\{5, 6\}$ will be a null set $\{\ \}$ because no object belongs to both sets $\{1, 2, 3\}$ and $\{5, 6\}$.

The complement of a set. The complement of a set A denoted as $\overline{A}$, is the set of all those objects of the universal set which do not belong to A. The complement of a set A can be graphically represented by the Venn diagram Figure 2.3.

For example, in rolling a dice experiment, the universal set Ω is $\Omega = \{1, 2, 3, 4, 5, 6\}$. The complement $\overline{A}$ of a set $A = \{1, 2\}$ is $\overline{A} = \{3, 4, 5, 6\}$. It is obvious that the union of a set A and its complement set $\overline{A}$ will always be the universal set Ω, that is, $\Omega \equiv A \cup \overline{A}$.

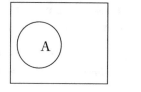
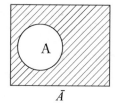

$\bar{A}$

Figure 2.3: Complement of set A.

The following are several rules about the operations of union, interaction, and complement of sets.

Commutative rule: The union and interaction of sets are commutative, that is,

$$A \cup B = B \cup A \tag{2.1}$$

$$A \cap B = B \cap A. \tag{2.2}$$

Example 2.1

In rolling a dice experiment, sets A and B are $A = \{1, 2, 3, 4\}$ and $B = \{1, 4, 5\}$. Use these two sets to demonstrate Equations (2.1) and (2.2).

Solution:

According to the definition of the union of two sets, we have:

$$A \cup B = \{1, 2, 3, 4\} \cup \{1, 4, 5\} = \{1, 2, 3, 4, 5\}$$
$$B \cup A = \{1, 4, 5\} \cup \{1, 2, 3, 4\} = \{1, 2, 3, 4, 5\}.$$

Therefore, $A \cup B = B \cup A$.

According to the definition of the intersection of two sets, we have:

$$A \cap B = \{1, 2, 3, 4\} \cap \{1, 4, 5\} = \{1, 4\}$$
$$B \cap A = \{1, 4, 5\} \cap \{1, 2, 3, 4\} = \{1, 4\}.$$

Therefore, $A \cap B = B \cap A = \{1, 4\}$.   ∎

Associative rule: The union and intersection of sets are associative, that is,

$$(A \cup B) \cup C = A \cup (B \cup C) \tag{2.3}$$
$$(A \cap B) \cap C = A \cap (B \cap C). \tag{2.4}$$

Example 2.2

In rolling a dice experiment, sets A, B, and C are $A = \{1, 4\}$, $B = \{2, 4, 5\}$, and $C = \{3\}$. Use these three sets to demonstrate Equations (2.3) and (2.4).

Solution:

According to the definition of the union of two sets, we have:

$$(A \cup B) \cup C = (\{1, 4\} \cup \{2, 4, 5, \}) \cup \{3\} = \{1, 2, 4, 5, \} \cup \{3\} = \{1, 2, 3, 4, 5\}$$
$$A \cup (B \cup C) = \{1, 4\} \cup (\{2, 4, 5, \} \cup \{3\}) = \{1, 4\} \cup \{2, 3, 4, 5, \} = \{1, 2, 3, 4, 5\}.$$

Therefore, $(A \cup B) \cup C = A \cup (B \cup C) = \{1, 2, 3, 4, 5\}$.

According to the definition of the intersection of two sets, we have:

$$(A \cap B) \cap C = (\{1, 4\} \cap \{2, 4, 5, \}) \cap \{3\} = \{4\} \cap \{3\} = \emptyset$$
$$A \cap (B \cap C) = \{1, 4\} \cap (\{2, 4, 5, \} \cap \{3\}) = \{1, 4\} \cap \emptyset = \emptyset.$$

Therefore, $(A \cap B) \cap C = A \cap (B \cap C) = \emptyset$. ∎

Distributive rule: The union and intersection of sets are distributive.

$$(A \cup B) \cap C = (A \cap C) \cup (B \cap C) \tag{2.5}$$
$$(A \cap B) \cup C = (A \cup C) \cap (B \cup C). \tag{2.6}$$

Example 2.3

In rolling a dice experiment, sets A, B, and C are $A = \{2, 3\}$, $B = \{1, 4, 5\}$, and $C = \{1, 3, 6\}$. Use these three sets to demonstrate Equations (2.5) and (2.6).

Solution:

According to the definitions of union and intersection of two sets, we have:

$$(A \cup B) \cap C = (\{2, 3\} \cup \{1, 4, 5\}) \cap \{1, 3, 6\} = \{1, 2, 3, 4, 5\} \cap \{1, 3, 6\} = \{1, 3\}$$
$$(A \cap C) \cup (B \cap C) = (\{2, 3\} \cap \{1, 3, 6\}) \cup (\{1, 4, 5\} \cap \{1, 3, 6\}) = \{3\} \cup \{1\} = \{1, 3\}.$$

So, $(A \cup B) \cap C = (A \cap C) \cup (B \cap C) = \{1, 3\}$.

According to the definitions of union and intersection of two sets, we have:

$$(A \cap B) \cup C = (\{2, 3\} \cap \{1, 4, 5\}) \cup \{1, 3, 6\} = \emptyset \cup \{1, 3, 6\} = \{1, 3, 6\}$$
$$(A \cup C) \cap (B \cup C) = (\{2, 3\} \cup \{1, 3, 6\}) \cap (\{1, 4, 5\} \cup \{1, 3, 6\})$$
$$= \{1, 2, 3, 6\} \cap \{1, 3, 4, 5, 6\} = \{1, 3, 6\}.$$

Therefore, $(A \cap B) \cup C = (A \cup C) \cap (B \cup C) = \{1, 3, 6\}$. ∎

de Morgan's rule: This rule states that the complement of unions (intersections) is equal to the intersections (unions) of the respective complements, that is,

$$\overline{A \cup B} = \overline{A} \cap \overline{B} \tag{2.7}$$

$$\overline{A \cap B} = \overline{A} \cup \overline{B}. \tag{2.8}$$

Example 2.4

In rolling a dice experiment, sets A and B are $A = \{1, 4, 6\}$ and $B = \{2, 4, 5\}$. Use these two sets to demonstrate Equations (2.7) and (2.8).

Solution:

For rolling a dice experiment, the universal set is $\Omega = \{1, 2, 3, 4, 5, 6\}$.

According to the definition of the complement of a set, we have:

$$A = \{1, 4, 6\}, \overline{A} = \{2, 3, 5\}$$
$$B = \{2, 4, 5\}, \overline{B} = \{1, 3, 6\}.$$

According to the definition of the union of two sets, the complement of a set and intersection of two sets, we have:

$$\overline{A \cup B} = \overline{\{1, 4, 6\} \cup \{2, 4, 5\}} = \overline{\{1, 2, 4, 5, 6\}} = \{3\}$$
$$\overline{A} \cap \overline{B} = \{2, 3, 5\} \cap \{1, 3, 6\} = \{3\}.$$

Therefore, $\overline{A \cup B} = \overline{A} \cap \overline{B} = \{3\}$.

According to the definition of the intersection of two sets, the complement of a set and union of two sets, we have:

$$\overline{A \cap B} = \overline{\{1, 4, 6\} \cap \{2, 4, 5\}} = \overline{\{4\}} = \{1, 2, 3, 5, 6\}$$
$$\overline{A} \cup \overline{B} = \{2, 3, 5\} \cup \{1, 3, 6\} = \{1, 2, 3, 5, 6\}.$$

Therefore, $\overline{A \cap B} = \overline{A} \cup \overline{B} = \{1, 2, 3, 5, 6\}$.   ■

2.4   DEFINITION OF PROBABILITY

Probability is the measure of a likelihood that an event will occur. The calculation of probability can be based on two typical definitions: the relative frequency and the axiomatic definition.

2.4.1   RELATIVE FREQUENCY

One classical approach to define a probability is the relative frequency. The probability of the occurrence of an event E denoted as $P(E)$ is defined as the ratio of the number of occurrences of the event A to the total number of trials.

$$P(E) = \lim_{N \to \infty} \left(\frac{n}{N} \right), \tag{2.9}$$

where n is the number of occurrences of the event E and N is the total number of trials of the experiment. Since minimum and maximum possible values of n are 0 and N, we have:

$$0 \le P(E) \le 1. \tag{2.10}$$

This definition gives us a tool to calculate the probability of an event if a set of data of trials are given.

Example 2.5
One person has tossed a coin 1,000 times, got 482 "heads" and 518 "tails." Calculate the probability of showing "head" and the probability of showing "tail."

Solution:
Let A represent the event of showing "head" and B the event of showing "tail." So,

$$A = \{head\}, \qquad B = \{tail\}.$$

According to the relative frequency definition of probability, we have:

$$P(A) = \frac{482}{100} = 0.482, \qquad P(B) = \frac{518}{1000} = 0.518.$$

■

Example 2.6
A company designs and distributes a machine unit, which has a specified service life of two years. The technical data shows that a total of 25,192 units have been sold and 1,053 units failed during the two-year service. Estimate the probability of the failed unit and the safe unit during the two-year service.

Solution:
Let A represent the event of failure during two-year service and B the event of working safely during two-year service. So,

$$A = \{Failure\}, \qquad B = \{Safe\}.$$

According to the relative frequency definition of probability, the probability of the failed machine unit will be:

$$P(A) = P(failure) = \frac{1053}{25,192} = 0.042 = 4.2\%.$$

The probability of the safe machine unit will be:

$$P(B) = \frac{25{,}192 - 1053}{25{,}192} = 0.958 = 95.8\%.$$

∎

2.4.2   AXIOMATIC DEFINITION

In the axiomatic approach to defining probability, it is assumed that the occurrence of every sample point in an experiment will have the same likelihood. For an example of tossing a coin, the occurrence of the likelihood of "head" or "tail" is the same. For another example, the occurrence of showing number 1 in rolling a dice will have the same likelihood as the occurrence of showing number 6.

According to the axiomatic approach, the probability of occurrence of an event E, denoted as $P(E)$, is defined as a numerical value of $P(E)$ such that $P(E)$ obeys the following three axioms or postulates.

1. The probability of an event E cannot be a negative value because it has not any physical meaning. The event E can happen with a positive probability or cannot happen with a zero probability, that is,

$$P(E) \geq 0. \tag{2.11}$$

2. If E is a universal set Ω, the occurrence of a universal set Ω is a certain event, that is

$$P(\Omega) = 1. \tag{2.12}$$

3. If events E_1 and E_2 are mutually exclusive, the probability of the union of two events is equal to the sum of the probability of each mutually exclusive event, that is

$$P(E_1 \cup E_2) = P(E_1) + P(E_2). \tag{2.13}$$

Example 2.7

The experiment of rolling a dice will have six possible sample point or outcome, that is, number 1, 2, 3, 4, 5, or 6. (1) Calculate the probability of showing number 5. (2) If the set $A = \{3, 5\}$, calculate the probability of the set A.

Solution:

1. Calculate the probability of showing number 5.

The universal set of rolling dice is $\Omega = \{1, 2, 3, 4, 5, 6\}$. There are six sample points in the universal set. Since all of the sample points are a mutually exclusive event, we have:

$$P(\Omega) = 1 = P(\{1, 2, 3, 4, 5, 6\})$$
$$= P(\{1\}) + P(\{2\}) + P(\{3\}) + P(\{4\}) + P(\{5\}) + P(\{6\}).$$

Since each sample point will have the same occurrence likelihood, we have

$$P(\{1\}) = P(\{2\}) = P(\{3\}) = P(\{4\}) = P(\{5\}) = P(\{6\}).$$

Therefore, $P(\{1\}) = P(\{2\}) = P(\{3\}) = P(\{4\}) = P(\{5\}) = P(\{6\}) = \dfrac{1}{6}$.

2. $A = \{3, 5\}$, calculate $P(A)$.

$$P(A) = P(\{3\}) + P(\{5\}) = \frac{1}{6} + \frac{1}{6} = \frac{1}{3}.$$

∎

Example 2.8

There are three white balls and four red balls in a box. The experiment of randomly picking one ball from the box is conducted for a total of 500 times. Among these 500 experiments, we get 203 white balls and 397 red balls. (1) Use the relative frequency definition to calculate the probability of picking a white ball. (2) Use the axiomatic definition to calculate the probability of picking a white ball.

Solution:

- P (*whiteball*) by the relative frequency definition.

  According to the relative frequency definition, we have

  $$P(whiteball) = \frac{203}{500} = 0.406 = 40.6\%.$$

- P (*whiteball*) by the axiomatic definition.

  According to the axiomatic definition, we can have

  $$P(whiteball) = \frac{3}{7} = 0.429 = 42.9\%.$$

There is some difference between the two results obtained above. When we calculate the probability of an event by the axiomatic definition, we do not need to do any actual experiment. We need to calculate a number of all possible sample points (outcomes) and run the corresponding calculations based on that all sample points have the exact occurrence likelihood. In this example, we have a total of 7 possible sample points, and 3 of them are the occurrence of a white ball. So, $P\ (whiteball) = \frac{3}{7}$. However, the probability by the relative frequency definition will depend on the total number of experiments. When the total number of the experiment is toward an infinite, the probability calculated by the relative frequency definition will be equal to that by the axiomatic definition. ∎

Example 2.9

For the experiment of rolling two dice, calculate the probability of the sum of two numbers less than 5.

Solution:

For the experiment of rolling 2 dice, there is a total of 36 possible sample points (outcomes). The sum of two numbers less than 5 will only include following 6 sample points: $\{1, 3\}$, $\{3, 1\}$, $\{2, 2\}$, $\{1, 2\}$, $\{2, 1\}$, and $\{1, 1\}$. Therefore, the probability of the sum of two numbers less than 5 will be:

$$P\ (sum\ less\ than\ 5) = \frac{6}{36} = \frac{1}{6}.$$

∎

2.5 SOME BASIC OPERATIONS OF PROBABILITY

Based on the axiomatic definition of probability and the set theory, we have the following simple operations of the probability.

2.5.1 PROBABILITY OF MUTUALLY EXCLUSIVE EVENTS

When the two sets (events) E_1 and E_2 are mutually exclusive events, that is, no common elements as shown in Figure 2.4, the probability $P\ (E_1 \cap E_2)$ of the intersection of mutually exclusive events will be zero:

$$P\ (E_1\cap) = P\ (\emptyset) = 0. \tag{2.14}$$

The probability of two mutually exclusive events will be:

$$P\ (E_1 \cup E_2) = P\ (E_1) + P\ (E_2). \tag{2.15}$$

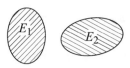

Figure 2.4: The Venn diagram of two mutually exclusive events.

Example 2.10
In a class survey, the percentage of five options: "Strongly agree," "Agree," "No opinion," "Disagree," and "Strongly disagree" on a survey question are 24%, 48%, 8%, 15%, and 5%. What is the percentage of the answers with at least "Agree" on the survey question?

Solution:
Since the option of "Strongly agree" and "Agree" are a mutually exclusive event, we have:

$$P\left(\text{"Strongly agree"} \cup \text{"Agree"}\right) = P\left(\text{"strongly agree"}\right) + P\left(\text{"Agree"}\right)$$

$$= 0.24 + 0.48 = 0.72.$$

∎

2.5.2   PROBABILITY OF AN EVENT IN A FINITE SAMPLE SPACE Ω

Assume that event A in a finite sample space Ω contain n sample points (outcomes), that is, $A = \{E_1, \ldots, E_n\}$. Since each sample point is a mutually exclusive event with any other sample point (outcome), the probability of an event A in a finite sample space Ω is

$$P\left(A\right) = P\left(\{E_1, \ldots, E_n\}\right) = \sum_{i=1}^{i=n} P\left(E_i\right). \tag{2.16}$$

Example 2.11
In an experiment of rolling a dice, calculate the probability of event $A = \{1, 4, 5\}$.

Solution:
Since the event A contains three sample points, the probability of each sample point in rolling a dice experiment will be 1/6. So the probability $P\left(A\right)$ is

$$P\left(A\right) = P\left(\{1, 4, 5\}\right) = P\left(\{1\}\right) + P\left(\{4\}\right) + P\left(\{5\}\right)$$

$$= \frac{1}{6} + \frac{1}{6} + \frac{1}{6} = \frac{1}{3}.$$

∎

Example 2.12
The outcome of an experiment of rolling a dice twice will be the sum of showing numbers. Calculate the probability of an event $A = \{4, 7\}$.

Solution:

In this experiment, we will have a total $6 \times 6 = 36$ possible sample points. The event $B_1 = \{4\}$ contains three possible outcomes: $(1+3), (2+2)$, and $(3+1)$, so the probability of $P(\{4\})$ will be

$$P(\{4\}) = \frac{3}{36}.$$

In the same approach, the event $B_2 = \{7\}$ will contain six possible outcomes: $(1+6), (2+5), (3+4), (4+3), (5+2)$, and $(6+1)$. The probability of $P(\{7\})$ will be:

$$P(\{7\}) = \frac{6}{36}.$$

Therefore, according to Equation (2.16), we have:

$$P(A) = P(\{4,7\}) = P(\{4\}) + P(\{7\})$$

$$= \frac{3}{36} + \frac{12}{36} = \frac{5}{12}.$$

■

2.5.3   PROBABILITY OF UNION AND INTERSECTION OF TWO EVENTS

If events E_1 and E_2 are two events (sets) of the same experiment, the probability of the union of two events is:

$$P(E_1 \cup E_2) = P(E_1) + P(E_2) - P(E_1 \cap E_2). \tag{2.17}$$

Figure 2.5 clearly shows that the doubled crossed area is the intersection of the event E_1 and E_2. In the probability of $P(E_1) + P(E_2)$, the doubled crossed area is counted twice.

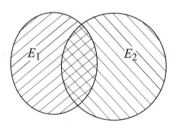

Figure 2.5: The Venn diagram of the union of the event E_1 and E_2.

We can rearrange Equation (2.17) to form the formula to calculate the probability of the intersection of two events E_1 and E_2 as,

$$P(E_1 \cap E_2) = P(E_1) + P(E_2) - P(E_1 \cup E_2). \tag{2.18}$$

Example 2.13
Among the graduate students of a university, 42% receive loans, 25% receive teaching assistantships, and 14% receive both. What % of students receive either a loan or an assistantship?

Solution:
Let the events A and B represent receiving a loan and research assistantships, respectively. According to the given information, we have:

$$P(A) = 0.42; \quad P(B) = 0.25 \quad \text{and} \quad P(A \cap B) = 0.14.$$

Therefore, the probability of students receives either a loan or an assistantship will be:

$$P(A \cup B) = P(A) + P(B) - P(A \cap B) = 0.42 + 0.25 - 0.14 = 0.53 = 53\%.$$

∎

Example 2.14
A student could select a science course denoted as event A; a math course denoted as event B or none of these two courses. From the registering information, 40% of students have chosen a science course, 50% a math course, and 75% at least one of a science and a math course. Calculate the percentage of students who select both a science and a math course.

Solution:
According to the given information, we have:

$$P(A) = 0.40; \quad P(B) = 0.50 \quad \text{and} \quad P(A \cup B) = 0.75.$$

According to Equation (2.15), we have

$$P(A \cap B) = P(A) + P(B) - P(A \cup B) = 0.40 + 0.50 - 0.75 = 0.15.$$

∎

2.5.4   PROBABILITY OF A COMPLEMENTARY EVENT
Assume that the probability of an event A in a sample space Ω is $P(A)$, then the probability of the complementary event $\overline{A}$ is:

$$P(\overline{A}) = 1 - P(A). \tag{2.19}$$

Example 2.15
The outcome of an experiment of rolling a dice twice will be the sum of showing numbers. Calculate the probability of the sum showing numbers not equal to 5.

Solution:

Let A represent the event of the sum of the showing number equal to 5, that is, $A = \{5\}$. This event A contains four possible sample points: $(1 + 4)$, $(2 + 3)$, $(3 + 2)$, and $(4 + 1)$. The probability of $P(\{5\})$ will be:

$$P(\{5\}) = \frac{4}{36}.$$

Then, the probability of the sum of the showing numbers not equal to 5 is

$$P(\overline{A}) = 1 - P(A) = 1 - \frac{4}{36} = \frac{8}{9}.$$

∎

2.5.5   PROBABILITY OF STATISTICALLY INDEPENDENT EVENTS

Statistically independent events: If the occurrence of the event E_1 in no way affects the probability of occurrence of the event E_2, the events E_1 and E_2 are statistically independent events. An example of two statistically independent events are rolling dice and tossing a coin.

When two events E_1 and E_2 are statistically independent, the intersection of events E_1 and E_2 just means that both occur. The probability of the intersection of two statistically independent event E_1 and E_2 will be equal to the multiplication of probabilities of each event, that is,

$$P(E_1 \cap E_2) = P(E_1) \times P(E_2). \tag{2.20}$$

Example 2.16

There is a total of seven balls in a box including three red balls and four green balls. Randomly pick one ball from the box, record the color, and put it back. Calculate the probability of the red ball in the first picking and the blue ball in the second picking.

Solution:

Let event E_1 to represent the red ball in the first picking and event E_2 the blue ball in the second picking.

The probability of the event E_1, that is, red ball in first picking will be:

$$P(E_1) = \frac{3}{7}.$$

The probability of the event E_2, that is, blue ball in second picking will be:

$$P(E_2) = \frac{4}{7}.$$

In this problem, the ball will be put back into the box. Therefore, these two events E_1 and E_2 are statistically independent. So, according to Equation (2.20), the probability of the red ball in the first picking and the blue ball in the second picking is

$$P(E_1 \cap E_2) = P(E_1) \times P(E_2) = \frac{3}{7} \times \frac{4}{7} = \frac{12}{49}.$$

Example 2.17
A water pump is driven by an electric motor. The reliability of the water pump is 0.95. The reliability of the electrical motor is 0.99. Calculate the reliability of this unit, including an electric motor and water pump.

Solution:
Let event E_1 to represent that the electric motor works properly and event E_2 that the water pump properly works. According to the given information, we have:

$$P(E_1) = 0.99, \qquad P(E_2) = 0.95.$$

The electrical motor and the water pump are two different products. They function according to their working principles. So, it is obvious that E_1 and E_2 are statistically independent. Therefore, the probability of the unit, including the electric motor and the water pump is:

$$P(E_1 \cap E_2) = P(E_1) \times P(E_2) = 0.99 \times 0.95 = 0.94.$$

∎

2.5.6   CONDITIONAL PROBABILITY

The occurrence of the event E_2 when the event E_1 has occurred is known as the conditional event and is denoted as $E_2 | E_1$. The conditional probability of this conditional event $E_2 | E_1$ is defined as

$$P(E_2 | E_1) = \frac{P(E_2 \cap E_1)}{P(E_1)} \qquad P(E_1) > 0. \tag{2.21}$$

Equation (2.21) can also be expressed as

$$P(E_2 \cap E_1) = P(E_2 | E_1) P(E_1) = P(E_1 | E_2) P(E_2). \tag{2.22}$$

If these two events E_1 and E_2 are statistically independent, we can have

$$P(E_2 | E_1) = P(E_2). \tag{2.23}$$

Example 2.18
An engineering class has 67 students. Twenty-seven students choose a mechanical-concentrated program, 16 students choose an electrical-concentrated program, and the remaining 24 students

Table 2.1: The data for enrollments of two classes

	Event M	Event E	Event G
Total	27	16	24
Event A	10	8	18
Event B	17	8	6

a general engineering program. The data about the registration of a science class, an engineering class, and corresponding event symbols are shown in Table 2.1. Event M is mechanical-concentrated; event E is electrical-concentrated; event G is general engineering. Event A is a science class, and event B is an engineering class. Calculate $P(M \cap A)$

Solution:
According to the provided information, we have:

$$P(M) = \frac{27}{67}, \qquad P(A) = \frac{10 + 8 + 18}{67} = \frac{36}{67}.$$

According to the definition of the conditional probability and the provided information, we have:

$$P(A|M) = \frac{10}{27}, \qquad P(M|A) = \frac{10}{10 + 8 + 18} = \frac{10}{36}.$$

From Equation (2.22), we have

$$P(M \cap A) = P(A|M)P(M) = \frac{10}{27} \times \frac{27}{67} = \frac{10}{67}$$

$$P(M \cap A) = P(M|A)P(A) = \frac{10}{36} \times \frac{36}{67} = \frac{10}{67}.$$

■

Example 2.19
In a company, a well-planned design project is denoted as event A, and a well-executed project is denoted as event B. If the probability that a design project will be well planned is 0.75, and the probability that the design project will be well planned and well executed is 0.70. Calculate the probability of a well-planned design project that will also be well-executed, that is, $P(B|A)$.

Solution:
According to the given information, we have:

$$P(A) = 0.75, \qquad P(B \cap A) = 0.70.$$

From Equation (2.21), we have

$$P(B\,|A) = \frac{P\,(B \cap A)}{P\,(A)} = \frac{0.70}{0.75} = 0.933.$$

■

Example 2.20

In an experiment of rolling a dice twice, calculate the probability of showing number 4 in the second rolling when the number 5 has been shown in the first rolling.

Solution:

Let us use event A to represent showing number 4 in the first rolling and event B showing number 5 in the second rolling. Since event A and event B, in this case, are statistically independent, so we have

$$P(B\,|A) = P\,(B) = 1/6.$$

We can also use Equation (2.22) to run the calculation. In the first rolling, we can have $P\,(A) = \frac{1}{6}$. In rolling a dice twice, we can have a total 36 possible outcomes and showing number 5 and then 4 is one outcome, so we have $P\,(A \cap B) = \frac{1}{36}$. Therefore, we have:

$$P(B\,|A) = \frac{P\,(B \cap A)}{P\,(A)} = \frac{1/36}{1/6} = 1/6.$$

■

2.5.7 TOTAL PROBABILITY THEOREM

If $B_1, B_2, \ldots, B_n$ are a set of mutually exclusive, exhaustive collective event, that is,

$$B_j \cap B_i = \emptyset; \qquad j \neq i, i, j = 1, 2, \ldots, n$$
$$\Omega = B_1 \cup B_2 \cup \ldots \cup B_n.$$

The probability of another event A can always be expanded as

$$P\,(A) = P\,(A \cap B_1) + P\,(A \cap B_2) + \ldots + P\,(A \cap B_n)$$
$$= \sum_{i=1}^{n} P\,(A \cap B_i) = \sum_{i=1}^{n} P\,(A\,|B_i) \times P\,(B_i). \qquad (2.24)$$

Example 2.21

There was a total of 50 students in one engineering class with 40 male students and 10 female students. In their final grades, 12.5% of male students had a grade "A" and 10.0% of female had a grade "A". Calculate the probability of students with a grade "A".

Solution:

Let us use event E to represent a student with a grade "A", and the events B_1 and B_2 to represent a male and a female student, respectively.

According to the given information, we have:

$$P(B_1) = \frac{40}{50} = 0.8; \quad P(E \mid B_1) = 0.125; \quad P(B_2) = \frac{10}{50} = 0.2; \quad P(E \mid B_2) = 0.10.$$

According to Equation (2.24), we have:

$$P(E) = P(E \mid B_1) \times P(B_1) + P(E \mid B_2) \times P(B_2)$$
$$= 0.8 \times 0.125 + 0.2 \times 0.1 = 0.12 = 12\%.$$

■

Example 2.22

A company purchases one type of ball bearing from three different suppliers: 45% from supplier B_1, 35% from supplier B_2, and 20% from supplier B_3. According to the information provided by the suppliers, the probability of qualified bearings from the suppliers B_1, B_2, and B_3 are 92%, 95%, and 98%, respectively. Calculate the probability of a qualified bearing purchased by the company.

Solution:

Let event Q to represent a qualified bearing. From the given information, we have

$$P(B_1) = 0.45; \quad P(Q \mid B_1) = 0.92$$
$$P(B_2) = 0.35; \quad P(G \mid B_2) = 0.95$$
$$P(B_3) = 0.20; \quad P(G \mid B_3) = 0.98.$$

According to Equation (2.24), the probability of a qualified bearing in the company will be:

$$P(G) = P(G \mid B_1) \times P(B_1) + P(G \mid B_2) \times P(B_2) + P(G \mid B_3) \times P(B_3)$$
$$= 0.45 \times 0.92 + 0.35 \times 0.95 + 0.20 \times 0.98 = 0.9425 = 94.25\%.$$

■

2.5.8 BAYES' RULE

Bayes' rule (alternatively Bayes' theorem or Bayes' law) describes the probability of an event, based on prior knowledge of conditions that might be related to the event.

The Bayes' rule is an alternative expression of the conditional probability. If A and B are two events, since $P(A \cap B) = P(B \cap A)$, we can obtain the following equation according to Equations (2.21) or (2.22) of the conditional probability,

$$P(A \cap B) = P(A|B)P(B) = P(B \cap A) = P(B|A)P(A).$$

Rearranging this equation, we can have:

$$P(B|A) = \frac{P(A|B) \times P(B)}{P(A)}. \tag{2.25}$$

In Equation (2.25), $P(B|A)$ is the probability of event B when event A has happened. In the Bayes' rule, the probability $P(B|A)$ at the left end of the formula is typically called as the posterior probability. The probability $P(A|B)$ at the right end of the formula is the probability of occurrence of the event A when the event B has occurrence and is typically called as the prior probability.

A general expression of the Bayes' rule can be described as follows.

If B_1, B_2,...,B_n are a set of mutually exclusive, exhaustive collective event, that is,

$$B_j \cap B_i = \emptyset; \qquad j \neq i, i, j = 1, 2, \ldots, n$$
$$\Omega = B_1 \cup B_2 \cup \ldots \cup B_n$$

the probability of $P(B_j|A)$ will be

$$P\left(B_j|A\right) = \frac{P\left(A|B_j\right) \times P\left(B_j\right)}{\sum_{i=1}^{i=n} P\left(B_i\right) \times P\left(A|B_i\right)}. \tag{2.26}$$

Example 2.23

In an engineering statics class, 73% of students passed this class. For this class, students have a chance for additional learning from tutoring offered by the school. For this class, 18% of students attend the additional tutoring section. It is noted that 95% of students who have attended the tutoring section will pass the class. Calculate the probability of a student with a passing grade who attended the tutoring section.

Solution:

Let us use event A to represent a student with a passing grade and event T to represent attending additional tutoring section.

From the giving information, we have:

$$P(T) = 0.18; \qquad P(A|T) = 0.95; \qquad P(A) = 0.73.$$

Then, the probability of a student with a passing grade who attended the tutoring section, that is, $P(T|A)$, will be:

$$P(T|A) = \frac{P(A|T)\,P(T)}{P(A)} = \frac{0.95 \times 0.18}{0.73} = 0.234 = 23.4\%.$$

∎

Example 2.24

A product in a company is produced on three assembly lines. Three assembly lines account for 15%, 35%, and 50% of the product. The percentage of defective product items produced is 5% for the first assembly line; 4% for the second assembly line; and 1% for the third assembly line. If a product item is chosen at random from the total product output and is found to be defective, what is the probability that it was produced by the third assembly line?

Solution:

Let events B_1, B_2, and B_3 to represent the first, second, and third assembly line. Event A represents the defective unit. According to the provided information, we have:

$$P(B_1) = 0.15; \qquad P(A|B_1) = 0.05; \qquad P(B_2) = 0.35; \qquad P(A|B_2) = 0.04;$$

$$P(B_3) = 0.50; \qquad P(A|B_3) = 0.01.$$

These are prior probability based on information in the past. Now, we have a defective unit, the probability of this defective unit, which is caused by the third assembly line is a posterior probability and can be calculated as:

$$
\begin{aligned}
P(B_3|A) &= \frac{P(A|B_3) \times P(B_3)}{\sum_{i=1}^{i=3} P(B_i) \times P(A|B_i)} \\
&= \frac{P(A|B_3)\,P(B_3)}{P(A|B_1)\,P(B_1) + P(A|B_2)\,P(B_2) + P(A|B_3)\,P(B_3)} \\
&= \frac{0.01 \times 0.50}{0.05 \times 0.15 + 0.05 \times 0.35 + 0.01 \times 0.50} = 0.189 = 18.9\%.
\end{aligned}
$$

∎

2.6   RANDOM VARIABLE

The sample point (outcome) of an experiment can be described by a statement such as "head" or "tail" in tossing a coin experiment or can be described by a numerical value such as showing the number in rolling a dice experiment.

A random variable is a variable that associates a unique numerical value with every outcome of an experiment. The value of a random variable will vary from trial to trial as the experiment is repeated. For example, the diameter of a machined shaft is a random variable. In this example, the diameter of a machined shaft is an experiment. Before the shaft has been completely machined, the actual value of the diameter of the shaft is unpredicted. For another example, if we use a variable with a value 0 and 1 to represent the occurrence of "tail" and "head" in tossing a coin experiment, this variable is also a random variable.

In mechanical design, almost all of the design parameters can be described or expressed by random variables. For example, material strengths such as yield strength, ultimate strength, and fatigue strength are random variables. The loadings such as axial loading, bending moment, and torsion are random variables too.

2.7   MEAN, STANDARD DEVIATION, AND COEFFICIENT OF VARIANCE

When a set of sample data of a random variable has been observed or collected, three typical statistical characteristics—mean, standard deviation, and coefficient of variance—can be used to describe or represent the sample data.

Mean is a measure of the central value of a random variable. It is also termed as the expected value, mathematical expectation, or average. The mean can be calculated by the following equation when a set of sampling data of a random variable is collected:

$$\mu_x = \sum_{i=1}^{i=n} x_i/n, \tag{2.27}$$

where x_i is the ith sampling data of the random variable x and n is the number of sampling data. μ_x is the mean of random variable x.

Since the random variable will inherently have different sample values, the variation of the random variable should be considered.

Standard deviation is a measure of variation or dispersion of a set of sampling data around its central value:

$$\sigma_x = \begin{cases} \sqrt{\sum_{i=1}^{n} (x_i - \mu_x)^2 /(n-1)} & n < 30 \\ \sqrt{\sum_{i=1}^{n} (x_i - \mu_x)^2 /n} & n \geq 30, \end{cases} \tag{2.28}$$

where σ_x represents the standard deviation of a random variable x. The rest of the symbols in Equation (2.28) are the same as those in Equation (2.27).

The coefficient of variance is a standardized nondimensional measure of variation or dispersion of a set of sampling data around its central value. It is also known as relative standard deviation and is a much better tool to measure or describe the degree of variation of a random variable. The following equation can calculate the coefficient of variance:

$$\gamma_x = \frac{\sigma_x}{\mu_x}, \tag{2.29}$$

where γ_x is the coefficient of variance of a random variable x. σ_x and μ_x are the standard deviation and the mean of a random variable x.

When the number of sample size is less than 30, the mean, standard deviation, and coefficient of variance are the only three statistical characteristics for describing/representing this random variable. The mean is the expected value of the random variable, that is, the true value of the random variable if the uncertainty related with this random variable is eliminated. Both the standard deviation and the coefficient of variance can describe the variation of the data. The bigger the standard deviation and coefficient of variance are, the bigger the data scatter of the random variable is. However, the standard deviation is a measure of the absolute variance of the data around its mean. The coefficient of variance is a relative variance of data around its mean per unit of the mean. For example, the random variable X has a mean 10 (psi) and standard deviation 5 (psi). The random variable Y has a mean 100 (psi) and standard deviation 5 (psi). For these two random variables, they have the same standard deviation 5 (psi), but the coefficients of variance of X and Y are 0.5 and 0.05. This result means that the random variable X has a much bigger relative variation compared with the random variable Y.

When the number of a sample size of a random variable is more than 30, it is worthy of finding a statistical-related function for describing/presenting it, which will be discussed in Section 2.9.

Microsoft Excel and MATLAB are two available tools for engineers and engineering students. We can use them to run the calculations of the mean and the standard deviation.

In Microsoft Excel, the function for calculating the mean of sampling data $x_1, x_2, \ldots, x_n$ is

$$\text{average} \quad (x_1, x_2, \ldots, x_n) \qquad \text{for calculating the mean.}$$

In Microsoft Excel, there are several functions available to calculate the standard deviation of sampling data $x_1, x_2, \ldots, x_n$:

$$STDEV.S\ (x_1, x_2, \ldots, x_n) \qquad n < 30 \qquad \text{for calculating the standard deviation}$$
$$STDEV.P\ (x_1, x_2, \ldots, x_n) \qquad n \geq 30 \qquad \text{for calculating the standard deviation.}$$

In MATLAB, if the sample data is stored in a matrix A, the command for the mean and the standard deviation are:

$$mean(A) \qquad \text{for calculating mean}$$
$$std(A) \qquad \text{for calculating standard deviation.}$$

Example 2.25
The diameters of machined shafts by two machine operators are listed in Table 2.2. Analyze data and draw some comments about these two operators.

Table 2.2: The diameter of machined shafts

Operator	The Diameter of the Machined Shaft (inch)
Operator A	1.013, 0.944, 1.045, 1.022, 1.008, 1.069, 0.989, 1.009, 1.084, 0.967
Operator B	1.011, 1.024, 1.023, 1.007, 1.011, 1.007, 1.027, 1.009, 1.015, 1.016

Solution:
The diameter of a machined shaft by each operator is a random variable. Per Equations (2.26), (2.27), and (2.28) or the functions/commands in Excel or MATLAB, the mean, standard deviation, and coefficient of variance for operator A and B are listed in Table 2.3.

Table 2.3: Mean, standard deviation, and coefficient of variance of diameters of machined shafts

Operator	Mean	Standard Deviation	Coefficient of Variance
Operator A	1.015″	0.0431	0.0425
Operator B	1.015″	0.0073	0.0072

From Table 2.3, the means of machined shafts by both operators are the same. The standard deviation of the diameter of machined shaft by operator A is much bigger than that by operator B. The coefficient of variance of diameter of machine shaft by operator A is much bigger than that by the operator B. Therefore, we could say that operator B has a much better skill in machining because the diameter of machined shaft by him has a much smaller variation.   ■

Example 2.26
Ultimate tensile strength of two different material's tensile test specimens lists in Table 2.4. Calculate the mean, standard deviation, coefficient of variance, and make some comments about them.

Table 2.4: Ultimate tensile strength of two materials

Material	Ultimate Tensile Strength (ksi)
Material A	34.3, 31.9, 30.6, 34.8, 25.6, 33.1, 32.5, 29.3, 30.6, 26.5
Material B	67.4, 71.4, 73.9, 81.3, 68.3, 71.7, 70.7, 63.3, 69.2, 64.8

Solution:

The ultimate tensile strength is a random variable. Per Equations (2.27), (2.28), and (2.29) or the functions/commands in Excel or MATLAB, the mean, standard deviation, and coefficient of variance of ultimate tensile strengths of two materials are listed in Table 2.5.

Table 2.5: The mean, standard deviation, and coefficient of variance of ultimate tensile strength

Ultimate Tensile Strength	Mean	Standard deviation	Coefficient of variance
Material A	30.9 (ksi)	3.080	0.100
Material B	70.2 (ksi)	5.056	0.072

From Table 2.5, the mean of the ultimate tensile strength of material B is much bigger than that of materials A. So the material B is much stronger than material A. The standard deviation of ultimate tensile strength of materials B is 5.056 ksi, bigger than that of the material A. It means that the absolute variation about its mean for material B is bigger. However, the coefficient of variance of material A is 0.100 and is bigger than that of the material B. This says that the relative variation of the ultimate tensile strength of material A is much bigger than that of material B.   ∎

2.8   HISTOGRAM

2.8.1   DEFINITION OF A HISTOGRAM

When the sample size is big, such as more than 30, we can still use the mean, standard deviation, and coefficient of variance to describe them, but we can dig out more useful information from such big set of data. One typical tool is to build the histogram based on the data.

A histogram is a display of statistical information that uses rectangles to show the frequency of data items in successive numerical intervals of equal size.

An example of a histogram is shown in Figure 2.6, which is a histogram of a material's ultimate tensile strength. In a histogram, the width of a rectangular bar represents the range value of a bin, and the height of the bar is the frequency of sample data inside the bin's range. The bin in a histogram can be considered as a "bucket" with a lower boundary and an upper boundary. After a set of sample data of a random variable x are provided, the minimum and the maximum of the samples can be determined as x_{min} and x_{max}. Then, the bin size and values of the bin ranges can be determined by the following equations:

$$\Delta x = \frac{(x_{max} - x_{min})}{J}, \qquad J \geq 6, \tag{2.30}$$

$$B_{j-L} = x_{min} + (j-1) \times \Delta x, \qquad j = 1, 2, \ldots, J, \tag{2.31}$$

$$B_{j-U} = x_{min} + j \times \Delta x, \qquad j = 1, 2, \ldots, J, \tag{2.32}$$

where J is the number of bins, which is typically larger than or equal to 6. Δx is the range of the bin values. The symbol j represents the jth bin. B_{j-L} and B_{j-U} are the lower boundary and the upper boundary of the jth bin.

Through all the sample data x_i, we can use the following recurrence formula to determine the values of frequency in each bin:

$$n_j = \begin{cases} n_j + 1 & \text{for } j = 1 \text{ if } B_{j-L} \leq x_i \leq B_{j-U} \\ n_j + 1 & \text{for } j \neq 1 \text{ if } B_{j-L} < x_i \leq B_{j-U}, \end{cases} \tag{2.33}$$

where n_j is the frequency in the jth bin, that is, the number of sample data in the jth bin. x_i is the ith collected sampling data.

The histogram is a very useful tool to help us to smoothly understand an important concept of the PDF of a random variable and help us to tentatively select the type of distribution for a random variable, which will be discussed in Sections 2.9 and 2.13.

Now, we will use one example to show how to create and to explain a histogram.

Example 2.27
A set of 238 ultimate tensile strength data of material is listed in Table 2.6. Create and explain the histogram of the data with 15 bins.

Solution:
According to the data in Table 2.6, the minimum and maximum of the sampling data are 48.4 (ksi) and 76.8 (ksi). The range of a bin value per Equation (2.30) with 15 of bins is:

$$\Delta x = \frac{(x_{max} - x_{min})}{K} = \frac{(76.8 - 48.4)}{15} = 1.893 \text{ (ksi)}.$$

Per Equations (2.31) and (2.32), we can get the lower boundary and upper boundary of each bin as shown in the second column of Table 2.7. Per Equation (2.33), we can determine the frequency in each bin as shown in the third column of Table 2.7.

The histogram based on the data in the second and third column of Table 2.7 is shown in Figure 2.6. The histogram clearly shows that the frequencies in each bin are different even though the width or the range of each bin is the same. For example, there is 48 of sampling data in the 7th bin, but only 13 of sampling data in the 4th bin.

When the frequency of each bin is divided by the total number of sampling data, in this case, 238, we can get the relative frequency of each bin as listed in the 4th column of Table 2.7.

Table 2.6: A set of ultimate tensile strength data

Ultimate Tensile Strength (ksi)
63.8, 69.3, 51.7, 65.2, 62.8, 55.8, 59.6, 62.9, 76.8, 73.4, 55.6, 74.5, 64.6, 61.2, 64.5, 59.6, 60.9, 69.9, 71.5, 67.5, 64.3, 58.3, 64.5, 68.5, 63.6, 65.9, 64.6, 59.1, 62.7, 58.1, 65.3, 55.5, 56.9, 58.0, 48.8, 67.6, 62.8, 58.2, 67.3, 54.1, 61.0, 60.4, 62.8, 62.8, 58.7, 61.3, 60.7, 64.1, 66.1, 66.2, 57.7, 61.8, 56.2, 56.7, 61.4, 68.0, 58.1, 63.0, 60.5, 66.3, 56.8, 61.6, 63.8, 66.2, 68.1, 61.8, 55.0, 58.3, 56.9, 71.6, 58.8, 64.7, 60.6, 65.3, 58.2, 55.4, 55.3, 63.5, 60.7, 60.6, 67.6, 62.7, 62.3, 68.3, 58.0, 64.4, 65.1, 60.4, 62.4, 56.4, 56.5, 61.9, 64.6, 72.6, 58.6, 62.3, 61.1, 53.1, 59.6, 53.7, 65.1, 57.6, 61.9, 59.1, 62.8, 58.9, 63.6, 64.6, 68.8, 60.6, 52.3, 57.8, 67.3, 56.8, 65.6, 62.0, 67.6, 53.0, 60.6, 56.3, 73.9, 65.0, 67.4, 56.9, 59.4, 60.3, 66.2, 60.3, 64.5, 52.6, 59.9, 57.9, 54.7, 63.6, 62.7, 61.6, 55.7, 66.3, 63.0, 60.2, 61.5, 60.3, 53.9, 60.2, 57.9, 57.2, 56.5, 59.2, 52.8, 65.6, 63.7, 61.4, 61.3, 58.0, 65.8, 60.9, 58.4, 67.3, 60.5, 58.9, 60.2, 57.8, 56.6, 72.3, 68.6, 62.8, 56.0, 57.7, 60.7, 64.9, 55.7, 51.4, 55.2, 62.9, 63.1, 63.4, 60.9, 62.2, 59.4, 65.2, 55.6, 63.4, 57.8, 60.0, 63.8, 65.9, 56.6, 66.9, 64.3, 61.2, 60.6, 60.5, 60.1, 61.5, 61.7, 65.0, 68.0, 63.5, 60.5, 64.1, 62.2, 57.0, 65.5, 62.8, 62.0, 63.7, 62.6, 57.4, 60.8, 60.8, 59.2, 69.7, 57.7, 59.4, 58.4, 56.4, 60.6, 60.3, 68.0, 60.4, 56.9, 68.3, 66.8, 60.5, 55.0, 59.5, 60.8, 62.6, 60.3, 63.4, 63.1, 56.1, 57.4, 58.3, 59.3, 60.1, 61.5, 48.4,

Table 2.7: Statistical data of a histogram

Bin#	Bin's range	Frequency	Relative Frequency	Relative-Density Frequency
1	48.400–50.293	2	0.0084	0.0044
2	50.293–52.187	2	0.0084	0.0044
3	52.187–54.080	7	0.0294	0.0155
4	54.080–55.973	13	0.0546	0.0288
5	55.973–57.867	28	0.1176	0.0621
6	57.867–59.760	31	0.1303	0.0688
7	59.760–61.653	48	0.2017	0.1065
8	61.653–63.547	35	0.1471	0.0777
9	63.547–65.440	30	0.1261	0.0666
10	65.440–67.333	17	0.0714	0.0377
11	67.333–69.227	14	0.0588	0.0311
12	69.227–71.120	3	0.0126	0.0067
13	71.120–73.013	4	0.0168	0.0089
14	73.013–74.907	3	0.0126	0.0067
15	74.907–76.800	1	0.0042	0.0022

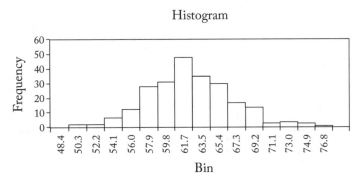

Figure 2.6: A histogram of a material's ultimate tensile strength.

Using the data in columns 2 and 4, we can create a relative-frequency bar chart, as shown in Figure 2.7. In this relative-frequency bar chart, the width of the bar is still the range of the bin, but the height of a bar represents the percentage (probability) of sampling data in the bin when they are compared with the whole set of sampling data. For example, there is a 20.17% of total 238 sampling data in the 7th bin, but only 0.84% of total 238 sampling data in the second bin.

When the relative frequency of each bin is divided by the width of the bin, in this case, 1.893, we can get a relative density frequency, which is the relative frequency per unit of the bin width. These values are listed in the fifth column of Table 2.7. Based on the second and the fifth column, we can create the relative-density frequency bar chart, as shown in Figure 2.8. One important aspect of the relative-density frequency bar chart is that the multiplication of the relative-density frequency (the bar height) with the range of the bin (the bin width) will be equal to the relative frequency (probability) of sampling data in this bin. For example, since the relative-density frequency in the fifth bin is 0.0621, and the width of this bin is 1.893, the probability (the relative frequency) of sampling data in this bin will be $0.0621 \times 1.893 = 0.1176$. If the sampling size is infinite, and the bin width is infinitesimal, the relative-density frequency will become a PDF, which is an extremely important concept and tool for describing random variable and will be discussed in detail in Section 2.9. ■

2.8.2   HISTOGRAM BY EXCEL AND MATLAB

When the sample size of the sample data of random variable is big, manual creation of histogram will be tedious. The histogram can be created by the tools in Excel and MATLAB.

In Excel, "Analysis ToolPak" is one of "add-in" program. After the "Analysis ToolPak" is added in, the tab "Data Analysis" will be available for creating a histogram as shown in Figure 2.9, which displays a list of functions including a histogram of the "Data Analysis" program.

One way to use the "Data Analysis" for creating a histogram is: (1) all of the sample data will be listed in a column in Excel; (2) using Equations (2.29), (2.30), and (2.31), we can

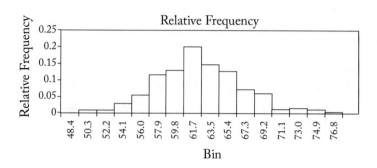

Figure 2.7: The relative-frequency bar chart of a material's ultimate tensile strength.

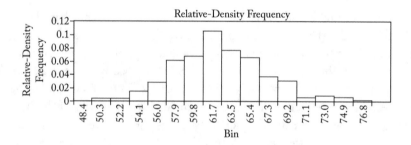

Figure 2.8: The relative-density frequency bar chart of a material's ultimate tensile strength.

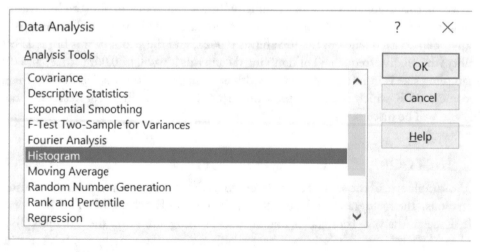

Figure 2.9: A list of functions of the "Data Analysis" program in Excel.

manually determine the lower boundary and upper boundary of each bin and arrange them in another column; and (3) activate the "Data Analysis," select "Histogram" and then follow the pop-up instructions in the prompt window. The "Data Analysis" will automatically count the frequencies in each bin per Equation (2.32) and create the histogram.

Histogram can be easily created in MATLAB by the following command:

$$\text{histogram} \quad (X, J),$$

where "X" is a matrix which contains all of the sample data of a random variable. "J" is the number of bins, which is typically larger than or equal to 6. Execution of the command "histogram (X, J)" will automatically create a histogram.

2.9   PROBABILITY FUNCTIONS

2.9.1   PROBABILITY FUNCTIONS OF A CONTINUOUS RANDOM VARIABLE

When the sample size of a continuous random variable is infinite, and the bin width is infinitesimal, the relative-density frequency, as shown in Figure 2.8 and discussed in the previous section will be a PDF.

Probability Density Function (PDF):   For a continuous random variable X, the function $f(x)$ is defined as a probability per unit of random variable value, named as PDF if it satisfies the following three conditions:

$$f(x) \geq 0 \qquad -\infty < x < \infty \tag{2.34}$$

$$P(a \leq X \leq b) = \int_a^b f(x)\, dx \tag{2.35}$$

$$P(-\infty \leq X \leq \infty) = \int_{-\infty}^{\infty} f(x)\, dx = 1. \tag{2.36}$$

Any function can be used as a PDF of a continuous random variable if it satisfies Equations (2.34), (2.35), and (2.36).

The physical meaning of $f(x)$ is the probability per unit of the random variable value, which is similar to the mass per unit length or the loading per unit length in beam theory. Since $f(x)$ is related to probability, a negative value of $f(x)$ has no physical meaning. Therefore, $f(x)$ must be large than or equal to 0, as shown in Equation (2.34).

Per the definition of $f(x)$, the probability of the event $(x \leq X \leq x + dx)$ is equal to $f(x)dx$, that is, $P(x \leq X \leq x + dx) = f(x)\, dx$. So, the probability of the event $(a \leq X \leq b)$ will be equal to $\int_a^b f(x)\, dx$, as shown in Equation (2.35).

Event $(-\infty \leq X \leq \infty)$ is a universal set of a continuous random variable. Therefore, the probability of the event $(-\infty \leq X \leq \infty)$ is certainly equal to 1, as shown in Equation (2.36).

We can use the PDF to define another important function: the cumulative distribution function.

Cumulative Distribution Function (CDF):   The CDF $F(x)$, also known as the PDF, of a random variable X is the probability that X will take a value less than or equal to x:

$$F(x) = P(-\infty \leq X \leq x) = \int_{-\infty}^{x} f(x)\, dx, \tag{2.37}$$

where $f(x)$ is the PDF of a random variable x.

If a random variable X is continuous and the first derivative of the CDF $F(x)$ exists, the PDF $f(x)$ is equal to the first derivative of $F(x)$, that is,

$$f(x) = \frac{dF(x)}{dx}. \tag{2.38}$$

Per the definition of the CDF as expressed in Equation (2.36), we can get following several conclusions about the CDF of a continuous random variable X:

$$F(-\infty) = \int_{-\infty}^{-\infty} f(x)\, dx = 0 \tag{2.39}$$

$$F(\infty) = \int_{-\infty}^{\infty} f(x)\, dx = 1 \tag{2.40}$$

$$P(a \leq X \leq b) = \int_{a}^{b} f(x)\, dx = \int_{-\infty}^{b} f(x)\, dx - \int_{-\infty}^{a} f(x)\, dx$$
$$= F(b) - F(a). \tag{2.41}$$

Example 2.28

The PDF $f(x)$ of a continuous random variable with a value in the range of $(-\infty, \infty)$ is shown in Figure 2.10. (1) Based on the curve, build the PDF of this random variable X. (2) Determine and plot the CDF. (3) Use both of the PDF and CDF to calculate the probability $P(1.5 \leq X \leq 3.5)$.

Solution:

1. The PDF of the random variable.

   Per the PDF curve provided, we can have the PDF of this random variable:

   $$f(x) = \begin{cases} 0 & -\infty \leq x \leq 1 \\ (x-1)/2 & 1 \leq x \leq 2 \\ (5-x)/6 & 2 \leq x \leq 5 \\ 0 & 5 \leq x \leq \infty. \end{cases}$$

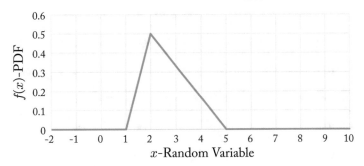

Figure 2.10: The curve of the PDF.

2. The CDF of the random variable.

Per Equation (2.36), the CDF of this random variable will be:

$$F(x) = \int_{-\infty}^{x} f(x)\,dx = \int_{-\infty}^{x} 0 \times dx = 0 \qquad -\infty \leq x \leq 1$$

$$F(x) = \int_{-\infty}^{x} f(x)\,dx = \int_{-\infty}^{1} f(x)\,dx + \int_{1}^{x} f(x)\,dx$$

$$= \int_{-\infty}^{1} 0 \times dx + \int_{1}^{x} \frac{1}{2}(x-1)\,dx = \frac{1}{2}\left(\frac{x^2}{2} - x + \frac{1}{2}\right) \qquad 1 \leq x \leq 2$$

$$F(x) = \int_{-\infty}^{x} f(x)\,dx = \int_{-\infty}^{1} f(x)\,dx + \int_{1}^{2} f(x)\,dx + \int_{2}^{x} f(x)\,dx$$

$$= \int_{-\infty}^{1} 0 \times dx + \int_{1}^{2} \frac{1}{2}(x-1)\,dx + \int_{2}^{x} \frac{1}{6}(5-x)\,dx$$

$$= 0.25 + \frac{1}{6}\left(5x - \frac{x^2}{2} - 8\right) \qquad 2 \leq x \leq 5$$

$$F(x) = \int_{-\infty}^{x} f(x)\,dx = \int_{-\infty}^{1} f(x)\,dx + \int_{1}^{2} f(x)\,dx + \int_{2}^{5} f(x)\,dx + \int_{5}^{x} f(x)\,dx$$

$$= \int_{-\infty}^{1} 0 \times dx + \int_{1}^{2} \frac{1}{2}(x-1)\,dx + \int_{2}^{5} \frac{1}{6}(5-x)\,dx + \int_{5}^{x} 0 \times dx$$

$$= 1 \qquad 5 \leq x \leq \infty.$$

The CDF of this random variable is plotted in Figure 2.11.

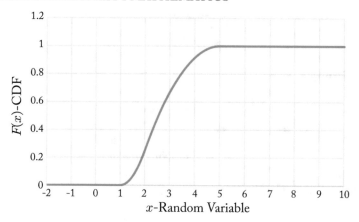

Figure 2.11: The curve of the CDF.

3. The probability $P\,(1.5 \leq X \leq 3.5)$.

The probability $P\,(1.5 \leq X \leq 3.5)$ can be determined by using the PDF $f\,(x)$ as follows:

$$P\,(1.5 \leq X \leq 3.5) = \int_{1.5}^{2} f\,(x)\,dx + \int_{2}^{3.5} f\,(x)\,dx$$

$$= \int_{1.5}^{2} \frac{1}{2}\,(x-1)\,dx + \int_{2}^{3.5} \frac{1}{6}\,(5-x)\,dx$$

$$= 0.1875 + 0.5625 = 0.75.$$

The probability $P\,(1.5 \leq X \leq 3.5)$ can also be directly determined by using the PDF $F(x)$ as follows:

$$P\,(1.5 \leq X \leq 3.5) = F\,(3.5) - F\,(1.5)$$

$$= \left[0.25 + \frac{1}{6}\left(5 \times 3.5 - \frac{3.5^2}{2} - 8\right)\right] - \left[\frac{1}{2}\left(\frac{1.5^2}{2} - 1.5 + \frac{1}{2}\right)\right]$$

$$= 0.8125 - 0.0625 = 0.75.$$

■

Example 2.29

The PDF of inner diameter X of a component in mass production is:

$$f\,(x) = \begin{cases} 0 & 1.000'' < x \\ 401e^{-401(x-1.000)} & x \geq 1.000''. \end{cases}$$

1. Determine the CDF.

2. Plot the PDF and the CDF.

3. If a component with an inner diameter larger than 1.010 will not be accepted, determine the percentage of the components that will be discarded.

Solution:

1. The CDF.

   Per Equation (2.37), we have the following.

   For $x < 1.000''$,

   $$F(x) = \int_0^x f(x)\,dx = \int_0^x 0 \times dx = 0.$$

   For $x \geq 1.000''$,

   $$F(x) = \int_0^x f(x)\,dx = \int_0^{1.000} f(x)\,dx + \int_{1.000}^x f(x)\,dx$$

   $$= \int_0^{1.000} 0 \times dx + \int_{1.000}^x 401e^{-401(x-1.000)}\,dx = 1 - e^{-401(x-1.000)}.$$

2. Plots of the PDF and the CDF.

   The plots of the PDF and the CDF are shown in Figures 2.12 and 2.13, respectively.

3. $P(x > 1.010'')$.

   Per the CDF obtained in step (1) of this example, the probability $P(x > 1.010'')$ is:

   $$P(x > 1.010) = 1 - P(x \leq 1.010) = 1 - F(1.010) = 1 - \left[1 - e^{-401(x-1.000)}\right]$$

   $$= 1 - 0.9819 = 0.0181 = 1.81\%.$$

   The meaning of the probability $P(x > 1.010'') = 1.81\%$ is that 1.81% of manufactured components will be rejected and discarded in this component mass production. ∎

2.9.2 PROBABILITY FUNCTIONS OF A DISCRETE RANDOM VARIABLE

For a discrete random variable such as rolling a dice, the PDF does not exist because there is a probability only at discrete points. In this case, we can use the PMF to describe the probability of a discrete random variable. The CDF is still applicable to a discrete random variable, but the formula is changed.

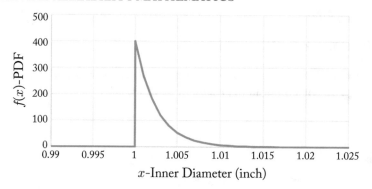

Figure 2.12: The PDF of the inner diameter.

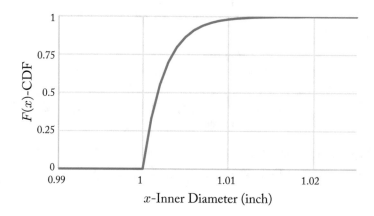

Figure 2.13: The CDF of the inner diameter.

Probability Mass Function (PMF): Probability mass function (PMF) of a discrete random variable X is a function, denoted as $p(x_i)$, that specifies the probability at a discrete point x_i:

$$p(x_i) = P(X = x_i),\qquad (2.42)$$

where $P(X = x_i)$ is the probability at the point $X = x_i$.

Cumulative distribution function (CDF): The CDF $F(x)$ of a discrete random variable X is the probability that X will take a value less than or equal to x:

$$F(x) = P(X \le x) = \sum_i p(x_i) \qquad x \le x_i,\qquad (2.43)$$

where $\sum_i p(x_i)$ is the sum of probabilities at the point where x_i is less than or equal to the value of x.

Example 2.30
Determine and plot the PMF and the CDF of a discrete random variable of rolling a dice.

Solution:
For rolling a dice, the discrete random variable X will have possible value $x = 1, 2, 3, 4, 5$, and 6. Since each side of a die is equally likely to show up, the probability of realizing any number between 1 and 6 is 1/6. So, the PMF with be:

$$p\,(1) = p\,(2) = p\,(3) = p\,(4) = p\,(5) = p\,(6) = \frac{1}{6}.$$

Per Equation (2.42), the CDF of rolling a dice will be:

$$F\,(x) = P\,(X < x) = \begin{cases} 0 & x < 1 \\ 1/6 & 1 \le x < 2 \\ 2/6 & 2 \le x < 3 \\ 3/6 & 3 \le x < 4 \\ 4/6 & 4 \le x < 5 \\ 5/6 & 5 \le x < 6 \\ 6/6 & 6 \le x. \end{cases}$$

The plots of the PMF and the CDF, in this case, are shown in Figures 2.14 and 2.15, respectively. It can be seen that the PMF is discrete dots because the PMF has only values at discrete points. At all other points, the PMF is zero. However, the CDF is step-curve. For example, the PMF is zero at $x = 4.5$, but the PDF at $x = 4.5$ is 4/6. ∎

2.10   MEAN OF A RANDOM VARIABLE

The PDF function for a continuous random variable and the PMF for a discrete random variable contain all the information about this random variable. Therefore, they can be used to determine the mean and standard deviation. Then, the coefficient of the variance can be calculated.

Mean, termed as the expected value, mathematical expectation, or average, has been defined in Section 2.7. It is a measure of the central value of a random variable and can be determined by the following equations.

For a continuous random variable X, the mean μ_x alternatively, $E(x)$ is

$$\mu_x = E\,(x) = \int_{-\infty}^{\infty} x f(x) dx, \tag{2.44}$$

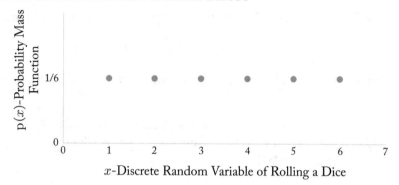

Figure 2.14: The plot of the PMF of rolling a dice.

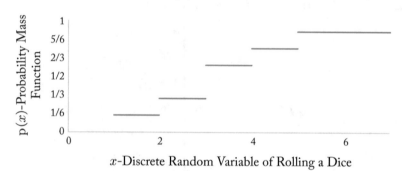

Figure 2.15: The plot of the CDF of rolling a dice.

where $f(x)$ is the PDF of a continuous random variable x.

For a discrete random variable X, the mean μ_x alternatively, $E(x)$ is

$$\mu_x = E(x) = \sum_{All\,i} x_i\, p(x_i), \tag{2.45}$$

where $p(x_i)$ is the PMF of a discrete random variable at the point x_i.

According to the definition of a random variable, any function $h(x)$ of a random variable x will be itself a random variable. The mean of $h(x)$ is defined as:

For a continuous random variable,

$$\mu_{h(x)} = E[h(x)] = \int_{-\infty}^{\infty} h(x) f(x)dx. \tag{2.46}$$

For a discrete random variable,

$$\mu_{h(x)} = E[h(x)] = \sum_{All\,i} h(x_i)\, p(x_i). \tag{2.47}$$

Example 2.31
Calculate the mean of a discrete random variable of rolling dice.

Solution:
For this discrete random variable X of rolling a dice, the PDF $p(x)$ is:

$$p(x) = \begin{cases} 1/6 & x = 1 \\ 1/6 & x = 2 \\ 1/6 & x = 3 \\ 1/6 & x = 4 \\ 1/6 & x = 5 \\ 1/6 & x = 6. \end{cases}$$

Per Equation (2.44), the mean of this discrete random variable of rolling dice is

$$\mu_x = E(x) = \sum_{All\ i} x_i\, p(x_i) = 1 \times \frac{1}{6} + 2 \times \frac{1}{6} + 3 \times \frac{1}{6} + 4 \times \frac{1}{6} + 5 \times \frac{1}{6} + 6 \times \frac{1}{6} = 3.5.$$

∎

Example 2.32
The PDF of the Brinell Hardness X of a component is

$$f(x) = \begin{cases} 0 & x < 300 \\ \dfrac{1}{70} & 300 \leq x \geq 370 \\ 0 & x > 370. \end{cases}$$

Determine the mean of the Brinell Hardness of the component.
Solution:
Per Equation (2.43), the mean of the Brinell Hardness of this component is:

$$\mu_x = E(x) = \int_0^\infty xf(x)\,dx = \int_0^{300} x \times 0\,dx + \int_{300}^{370} x\frac{1}{70}\,dx + \int_{370}^\infty x \times 0\,dx$$

$$= 0 + \frac{1}{140}x^2\Big|_{300}^{370} + 0 = 335 \ \text{(Hb)}.$$

∎

Example 2.33

The PDF of the diameter in millimeter of a shaft is:

$$f(x) = \begin{cases} 0 & x < 25.4 \text{ mm} \\ 18e^{-18(x-25.4)} & x \geq 25.4 \text{ mm}. \end{cases}$$

Calculate the mean of the shaft and its cross-section area.

Solution:

Per Equation (2.44), the mean of the shaft is:

$$\mu_x = E(x) = \int_{-\infty}^{\infty} xf(x)\,dx = \int_0^{25.4} x \times 0\,dx + \int_{25.4}^{\infty} x18e^{-18(x-25.4)}\,dx$$

$$= 0 + \left[-xe^{-18(x-25.4)} - \frac{1}{18}e^{-18(x-25.4)} \right]\Big|_{25.4}^{\infty} = 25.4 + \frac{1}{18} = 25.46 \text{ (mm)}.$$

The cross-section area of the shaft $A(x)$ is a function of the diameter, which is a random variable x, has the following function:

$$A(x) = \frac{\pi}{4}x^2.$$

Per Equation (2.46), the mean of the cross-section area of the shaft $A(x)$ is

$$\mu_{A(x)} = E[A(x)] = \int_{-\infty}^{\infty} A(x)\,f(x)\,dx$$

$$= \int_{-\infty}^{25.4} \frac{\pi}{4}x^2 \times 0\,dx + \int_{25.4}^{\infty} \frac{\pi}{4}x^2 18e^{-18(x-25.4)}\,dx$$

$$= 0 + \frac{\pi}{4} \left[-x^2 e^{-18(x-25.4)} - \frac{2}{18}xe^{-18(x-25.4)} - \frac{2}{18 \times 18}e^{-18(x-25.4)} \right]\Big|_{25.4}^{\infty}$$

$$= 508.93 \text{ (mm}^2\text{)}.$$

■

2.11   STANDARD DEVIATION AND COEFFICIENT OF VARIANCE

Before we can provide the formula for the standard deviation of a random variable, we need to define the variance of it first.

The variance of a random variable X is the mean of the function $(X - \mu_x)^2$ and is defined as

$$Var(X) = E\left[(X - \mu_x)^2\right] = E\left[X^2 - 2X\mu_x + \mu_x^2\right]$$

$$= E\left[X^2\right] - 2\mu_x E[X] + E\left[\mu_x^2\right] = E\left[X^2\right] - 2\mu_x^2 + \mu_x^2 = E\left[X^2\right] - \mu_x^2, \quad (2.48)$$

where $Var(X)$ refers to the variance of a random variable X and μ_x is the mean of a random variable X.

For a continuous random variable, the variance of a random variable X is

$$Var(X) = E[X^2] - \mu_x^2 = \int_{-\infty}^{\infty} x^2 f(x) \, dx - \mu_x^2. \tag{2.49}$$

For a discrete random variable, the variance of a random variable X is

$$Var(X) = E[X^2] - \mu_x^2 = \sum_{All\,i} x_i^2 p(x_i) - \mu_x^2. \tag{2.50}$$

In the previous section, Equation (2.28) is used to calculate the standard deviation of sampling data. After the definition of the PDF and the PMF are defined, the variance of a random variable can define the standard deviation.

Standard deviation is a measure of variation or dispersion of a set of data values around its central value and is defined as the square root of the variance of a random variable.

For a continuous random variable, the standard deviation of a random variable X is

$$\sigma_x = \sqrt{Var(X)} = \sqrt{E[X^2] - \mu_x^2} = \sqrt{\int_{-\infty}^{\infty} x^2 f(x) \, dx - \mu_x^2}. \tag{2.51}$$

For a discrete random variable, the standard deviation of a random variable X is

$$\sigma_x = \sqrt{Var(X)} = \sqrt{E[X^2] - \mu_x^2} = \sqrt{\sum_{All\,i} x_i^2 p(x_i) - \mu_x^2}, \tag{2.52}$$

where σ_x and $Var(X)$ are the standard deviation and the variance of the random variable X, respectively. $f(x)$ is the PDF of a continuous random variable X. $p(x_i)$ is the PMF of a discrete random variable X. μ_x is the mean of random variable X.

After the mean and standard deviation of a random variable X have been determined, we can define the coefficient of variance.

The coefficient of variance is the ratio of the standard deviation to the mean of a random variable and can be directly calculated by Equation (2.29), from page 44, which is repeated here:

$$\gamma_x = \frac{\sigma_x}{\mu_x}, \tag{2.29}$$

where γ_x is the coefficient of variance of a random variable x. σ_x and μ_x are the standard deviation and the mean of a random variable X.

Example 2.34
The PDF of the diameter in millimeter of a shaft is:

$$f(x) = \begin{cases} 0 & x < 20 \text{ mm} \\ 15e^{-15(x-20)} & x \geq 20 \text{ mm}. \end{cases}$$

Calculate its mean, standard deviation, and coefficient of variance.

Solution:
Per Equation (2.43), the mean of the shaft's diameter X is

$$\mu_x = E(X) = \int_{-\infty}^{\infty} x f(x) dx = \int_0^{20} x \times 0 dx + \int_{20}^{\infty} 15xe^{-15(x-20)} dx$$

$$= 0 + \left[-xe^{-15(x-20)} - \frac{1}{15} e^{-15(x-20)} \right] \Big|_{20}^{\infty} = 20 + \frac{1}{15} = 20.667 \text{ (mm)}$$

$$E(X^2) = \int_{-\infty}^{\infty} x^2 f(x) dx = \int_0^{20} x^2 \times 0 dx + \int_{20}^{\infty} 15x^2 e^{-15(x-20)} dx$$

$$= 0 + \left[-x^2 e^{-15(x-20)} - \frac{2}{15} xe^{-15(x-20)} - \frac{2}{15 \times 15} e^{-15(x-20)} \right] \Big|_{20}^{\infty}$$

$$= 20^2 + \frac{2}{15} 20 + \frac{2}{15 \times 15} = 402.676 \text{ (mm}^2).$$

Per Equation (2.50), the standard deviation of the shaft's diameter is

$$\sigma_x = \sqrt{Var(X)} = \sqrt{E[X^2] - \mu_x^2} = \sqrt{402.676 - 20.667^2} = 0.0667 \text{ (mm)}.$$

Per Equation (2.28), the coefficient of variance is:

$$\gamma_x = \frac{\sigma_x}{\mu_x} = \frac{0.0667}{20.0667} = 0.00332.$$

∎

Example 2.35
The number of automobiles arriving at a tollbooth per minute has the distribution in Table 2.8. Determine the mean, standard deviation, and coefficient of variance.

Table 2.8: Number of automobiles arriving at a tollbooth per minute

The number of automobiles, x	2	3	4	5	6	7	8	9
Probability mass function $p(x)$	0.06	0.09	0.15	0.22	0.18	0.16	0.10	0.04

Solution:
Per Equation (2.45), the mean of the number of automobiles arriving at a tollbooth per minute X (a discrete random variable) is:

$$\mu_x = E(X) = \sum_{All\, i} x_i\, p\,(x_i)$$

$$= 2 \times 0.06 + 3 \times 0.09 + 4 \times 0.15 + 5 \times 0.22$$

$$+ 6 \times 0.18 + 7 \times 0.16 + 8 \times 0.10 + 9 \times 0.04$$

$$= 5.45.$$

Per Equation (2.51), the standard deviation of a discrete random variable X is

$$\sigma_x = \sqrt{Var(X)} = \sqrt{E[X^2] - \mu_x^2} = \sqrt{\sum_{All\, i} x_i^2\, p\,(x_i) - \mu_x^2}$$

$$= \sqrt{\begin{array}{c} 2^2 \times 0.06 + 3^2 \times 0.09 + 4^2 \times 0.15 + 5^2 \times 0.22 + 6^2 \times 0.18 + 7^2 \times 0.16 \\ + 8^2 \times 0.10 + 9^2 \times 0.04 - 5.45^2 \end{array}}$$

$$= \sqrt{3.2075} = 1.7909.$$

Per Equation (2.28), the coefficient of variance of this discrete random variable X is

$$\gamma_x = \frac{\sigma_x}{\mu_x} = \frac{1.7909}{5.45} = 0.3286.$$

■

2.12   SOME TYPICAL PROBABILITY DISTRIBUTIONS

When the PDF or PMF of a random variable is specified and known, every possible information about it is fully defined and can be determined. Followings are several typical types of probability distributions for describing various types of a discrete and continuous random variable.

2.12.1   BINOMIAL DISTRIBUTION

Suppose that a random variable in one trial has only two outcomes. One outcome is named as "success" and represented by numerical number "1" with the probability p. Another outcome is named as "failure" and represented by a numerical number "0" with a probability q. Since this random variable has only two outcomes, we have:

$$p + q = 1. \tag{2.53}$$

The repeated independent n trials of such a random variable are called n-Bernoulli trials. The sample space of one such random variable has two sample points "1" as "success" and "0" as "failure." Then, the sample space of n-Bernoulli trials will have 2^n sampling points. For example, 3-Bernoulli trials will have $2^3 = 8$ sample points, which includes 000, 001, 010, 011, 100, 101, 110, and 111. Let X represents the number of successes in the n-Bernoulli trials. Then, the discrete random variable X of the n-Bernoulli trials can be described by the Binomial distribution.

Binomial distribution: For n-Bernoulli trials where a "success" is represented by a numerical number "1" with a probability p, the PMF of the random variable X in the n-Bernoulli trials with the number of "success" x follows the Binomial distribution expressed as:

$$p(X = x) = \binom{n}{x} p^x (1-p)^{n-x} \qquad \text{for } x = 0, 1, 2, \ldots, n, \tag{2.54}$$

where X is a discrete random variable which is equal to the number of "success" in the n-Bernoulli trials. x is the realizing value of the random variable X. $\binom{n}{x}$ is the number of possible combinations of x objects from a set of n objects and is equal to $\dfrac{n!}{x!(n-x)!}$. p is the probability of a "success" in one trial.

The CDF of the Binomial distribution per Equation (2.43) will be:

$$F(x) = P(X \leq x) = \sum_{k=0}^{k=x} \left[\binom{n}{k} p^{k n-k} \right]. \tag{2.55}$$

The Binomial distribution is fully defined by two distribution parameters n and p. The mean μ_X and standard deviation σ_X of a Binominal distribution X are determined by the following two equations:

$$\mu_X = E(X) = np \tag{2.56}$$

$$\sigma_X = \sqrt{var(X)} = \sqrt{np(1-p)}. \tag{2.57}$$

In Microsoft Excel, the functions for calculating PMF and the CDF of a Binomial distribution are:

$$p(x) = BINOM.DIST(x, n, p, FALSE) \tag{2.58}$$

$$F(x) = BINOM.DIST(x, n, p, TRUE). \tag{2.59}$$

In MATLAB, the commands for calculating the PMF and CDF of a binomial distribution are:

$$p(x) = binopdf(x, n, p) \tag{2.60}$$
$$F(x) = binocdf(x, n, p). \tag{2.61}$$

Example 2.36

A thermal power plant buys three boilers. It is assumed that the probability of a boiler functioning without failure for a year is 0.9. (1) Calculate and plot the PMF and CDF of this Binomial distribution. (2) Calculate the mean and standard deviation of it. (3) What will be the probability of an event with at least one boiler functioning without failure?

Solution:

1. The PMF and CDF.

   In this example, $n = 3$, $p = 0.9$. Per Equations (2.53), (2.57), or (2.59), we can calculate the PMF. Let use the function in Excel to run the calculations:

   $$p(x = 0) = BINOM.DIST(0, 3, 0.9, FALSE) = 0.001$$
   $$p(x = 1) = BINOM.DIST(1, 3, 0.9, FALSE) = 0.027$$
   $$p(x = 2) = BINOM.DIST(2, 3, 0.9, FALSE) = 0.243$$
   $$p(x = 3) = BINOM.DIST(3, 3, 0.9, FALSE) = 0.729.$$

   Per Equations (2.55), (2.59), or (2.61), we can calculate the CDF. Let use the command in MATLAB to run the calculations:

   $$F(0) = binocdf(0, 3, 0.9) = 0.001$$
   $$F(1) = binocdf(1, 3, 0.9) = 0.028$$
   $$F(2) = binocdf(2, 3, 0.9) = 0.271$$
   $$F(3) = binocdf(3, 3, 0.9) = 1.$$

   The PMF and CDF of this 3-Bernoulli trial are shown in Figures 2.16 and 2.17, respectively.

2. The mean and standard deviation.

   Per Equations (2.56) and (2.57), the mean and standard deviation are:

   $$\mu_X = E(X) = np = 3 \times 0.9 = 2.7$$

   $$\sigma_X = \sqrt{var(X)} = \sqrt{np} = \sqrt{3 \times 0.9 \times (1 - 0.9)} = 0.520.$$

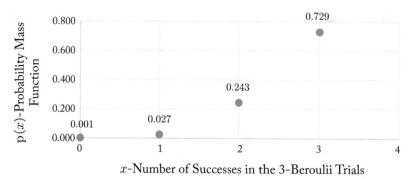

Figure 2.16: The PMFs of the Binomial distribution.

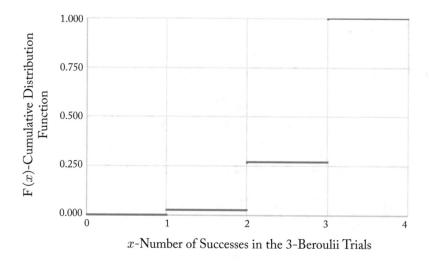

Figure 2.17: The CDF of the Binomial distribution.

3. The probability of an event with at least one boiler functioning without failure.

   The complimentary event of the event with at least one boiler functioning without failure is that all three boilers are in failure. Therefore,

$$P\left(X \geq 1\right) = 1 - p\left(x < 1\right) = 1 - P\left(x = 0\right) = 1 - 0.001 = 0.999.$$

■

2.12.2 POISSON DISTRIBUTION

In many engineering problems, the occurrence of an event is only affected by chance. Some examples are: (a) fatigue cracks that may occur in an automobile transmission shaft; (b) the number of telephone calls that may be received at any time over a specified period of time; and (c) the number of automobiles that may arrive at a tollbooth at any time over a specified period of time. The occurrence of such an event can be described by Poisson distribution.

Poisson Distribution: If an event occurs randomly and independently at any time or any point in space with the same likelihood at any subinterval, the random variable X, that denotes the number of events in an interval, can be described by the Poisson distribution. The PMF of Poisson distribution is

$$p\,(X = x) = \frac{e^{-\lambda}\lambda^x}{x!}; \qquad x = 0, 1, 2, \ldots, \tag{2.62}$$

where X is the number of occurrence events in the interval and x is the realizing value of the discrete random variable X. λ is the mean rate of occurrence of the event per the specified interval.

The Poisson distribution is fully specified by one distribution parameter λ. There are some notes about λ. For example, if the number of defects of a manufacturing bar per inch is 0.0001, the λ for a 20″-long bar will be:

$$\lambda = 0.0001 \times 20 = 0.002 \;\; \text{(defects per the 20″ bar)}.$$

In this example, the event is an occurrence of defects in a 20″-long bar, and the interval is not specified and is "present," that is, the present after the bars has been manufactured. The λ for a 5″-long bar will be

$$\lambda = 0.0001 \times 5 = 0.0005 \;\; \text{(defects per the 5″ bar)}.$$

For another example, if the average number of defects in a long underground cable per 1000-m length per year is 0.07, the λ for a 500-m length in a four year will be

$$\lambda = \frac{0.07}{1000 \times 1} \times (500 \times 4) = 0.14 \;\; \text{(defects per the 500-m length per 4 years)}.$$

In this example, the event is defects of the 500-m length cable in 4 years.

The CDF of a Poisson distribution per Equation (2.43) will be:

$$F\,(x) = P\,(X \le x) = \sum_{k=1}^{k=x} \frac{e^{-\lambda}\lambda^k}{k!}; \qquad x = 0, 1, 2, \ldots. \tag{2.63}$$

The mean and standard deviation of the Poisson distribution with a distribution parameter λ is

$$\mu_X = E\,(X) = \lambda \tag{2.64}$$

$$\sigma_X = \sqrt{var(X)} = \sqrt{\lambda}. \qquad (2.65)$$

In Microsoft Excel, the functions for calculating the PMF and CDF of a Poisson distribution are:

$$p(x) = POISSON.DIST(x, \lambda, FALSE) \qquad (2.66)$$
$$F(x) = POISSON.DIST(x, \lambda, TRUE). \qquad (2.67)$$

In MATLAB, the commands for calculating the PMF and CDF of a Poisson distribution are:

$$p(x) = poisspdf(x, \lambda) \qquad (2.68)$$
$$F(x) = poisscdf(x, \lambda). \qquad (2.69)$$

Example 2.37

The average number of defects in a long underground cable per 100-m length per year is 0.007. Calculate and tabulate the PMFs ($x = 0, 1, 2, 3, 4, 5$) of following two cases: (1) the number of defects in a cable with a length of 400 m in two years and (2) the number of defects in a cable with a length of 1500 m in 5 years.

Solution:

1. $p(x)$ for the number of defects in a cable with a length of 400 m in 2 years.

   The distribution parameter λ for the number of defects in a cable with a length of 400 m in a period two years is:

   $$\lambda = \frac{0.007}{100 \times 1} \times (400 \times 2) = 0.056 \text{ (defects per 400 m per 2 years)}.$$

   Per Equations (2.62), (2.65), or (2.67), we can calculate the PMF. Let us use Equation (2.65) to run the calculations:

   $$p(0) = POISSON.DIST(0, 0.056, FALSE) = 0.9455$$
   $$p(1) = POISSON.DIST(1, 0.056, FALSE) = 0.0530$$
   $$p(2) = POISSON.DIST(2, 0.056, FALSE) = 0.00148$$
   $$p(3) = POISSON.DIST(3, 0.056, FALSE) = 2.77 \times 10^{-5}$$
   $$p(4) = POISSON.DIST(4, 0.056, FALSE) = 3.87 \times 10^{-7}$$
   $$p(5) = POISSON.DIST(5, 0.056, FALSE) = 4.34 \times 10^{-9}.$$

2. $p(x)$ for the number of defects in a cable with a length of 1500 m in 5 years.

The distribution parameter λ for the number of defects in a cable with a length of 1500 m in 5 years is:

$$\lambda = \frac{0.007}{100 \times 1} \times (1500 \times 5) = 0.525 \quad \text{(defects per 1500 m per 5 years).}$$

Per Equations (2.62), (2.65), or (2.67), we can calculate the PMF. Let us use Equation (2.65) to run the calculations:

$$p\,(0) = poisspdf\,(0, 0.525) = 0.5916$$
$$p\,(1) = poisspdf\,(1, 0.525) = 0.3106$$
$$p\,(2) = poisspdf\,(2, 0.525) = 0.0815$$
$$p\,(3) = poisspdf\,(3, 0.525) = 0.0143$$
$$p\,(4) = poisspdf\,(4, 0.525) = 0.0019$$
$$p\,(5) = poisspdf\,(5, 0.525) = 0.0002.$$

The PMFs of these two Poisson distributions are tabulated in Table 2.9.   ■

Table 2.9: The PMFs of two Poisson distributions

x	0	1	2	3	4	5
$p(x)$ with $\lambda = 0.056$	0.9455	0.0530	0.00148	2.77E-5	3.87E-7	4.34E-9
$p(x)$ with $\lambda = 0.525$	0.5916	0.3106	0.0815	0.0143	0.0019	0.0002

2.12.3   UNIFORM DISTRIBUTION

Uniform distribution: If a random variable X has the same occurrence likelihood between the lower boundary a and the upper boundary b, it is a uniform distribution and often abbreviated as $X = U(a, b)$. The PDF of such uniform distribution with distribution parameters a and b is:

$$f\,(x) = \begin{cases} \dfrac{1}{b - a} & \text{for } a \le x \le b \\ 0 & \text{elsewhere.} \end{cases} \tag{2.70}$$

The CDF of a uniform distribution with distribution parameters a and b is:

$$F\,(x) = \begin{cases} 0 & \text{for } x < a \\ \dfrac{x - a}{b - a} & \text{for } a \le x \le b \\ 1 & \text{for } b < x. \end{cases} \tag{2.71}$$

The graphs of the PDF and CDF of a uniform distribution with distribution parameters a and b are shown in Figures 2.18 and 2.19.

The mean and standard deviation of a uniform distribution with distribution parameters a and b are

$$\mu_X = E\left(X\right) = \frac{a+b}{2} \tag{2.72}$$

$$\sigma_X = \frac{b-a}{\sqrt{12}}. \tag{2.73}$$

∎

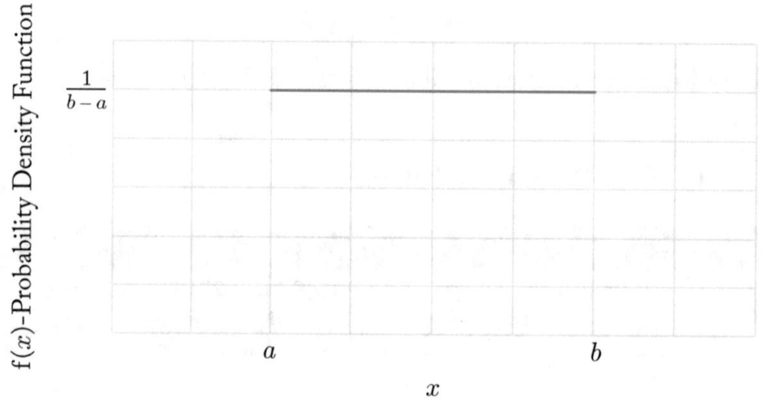

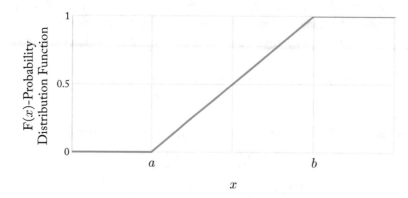

Figure 2.18: The graph of the PDF of a uniform distribution.

Figure 2.19: The graph of the CDF of a uniform distribution.

Example 2.38

The temperature in a device is uniformly distributed between 200°C and 221°C. (1) Build the PDF and CDF of this uniform distribution. (2) Calculate the probability $P(T \leq 215°C)$.

Solution:

1. The PDF and CDF of this uniform distribution.

   In this example, the distribution parameters are $a = 200$ and $b = 221$. Per Equations (2.69) and (2.70), the PDF and CDF are:

$$f(t) = \begin{cases} \dfrac{1}{221 - 200} = \dfrac{1}{21} & \text{for } 200 \leq t \leq 221 \\ 0 & \text{elsewhere.} \end{cases}$$

$$F(t) = \begin{cases} 0 & \text{for } t < 200 \\ \dfrac{t - 200}{21} & \text{for } 200 \leq t \leq 221 \\ 1 & \text{for } 221 < t. \end{cases}$$

2. Calculate the probability $P(T \leq 215°C)$.

   We can use the definition of the PDF to calculate the probability $P(T \leq 215°C)$.

$$P(T \leq 215) = \int_{-\infty}^{215} f(t)\, dt = \int_{-\infty}^{200} f(t)\, dt + \int_{200}^{215} f(t)\, dt$$

$$= \int_{-\infty}^{200} 0\, dt + \int_{200}^{215} \frac{1}{21}\, dt = 0 + \frac{t}{21}\Big|_{200}^{221} = \frac{215 - 200}{21} = 0.7143.$$

   $P(T \leq 215)$ can also directly calculated by using the CDF.

$$P(T \leq 215) = F(215) = \frac{215 - 200}{21} = 0.7143.$$

∎

2.12.4 NORMAL DISTRIBUTION

The normal distribution is a very common continuous probability distribution. It often describes the distribution of real-valued random variables whose distributions are not known and that are affected by lots of uncertainty.

Normal distribution: The PDF of a normally distributed random variable X, also known as a normal distribution or Gaussian distribution is:

$$f_X(x) = \frac{1}{\sqrt{2\pi}\sigma_x} \exp\left[-\frac{1}{2}\left(\frac{x - \mu_x}{\sigma_x}\right)^2\right] \qquad -\infty < x < \infty, \qquad (2.74)$$

where μ_x and σ_x are the mean and standard deviation of a normally distributed random variable, respectively.

The CDF of a normally distributed random variable X is

$$F_X(x) = P(X \leq x) = \int_{-\infty}^{x} \frac{1}{\sqrt{2\pi}\sigma_x} \exp\left[-\frac{1}{2}\left(\frac{x - \mu_x}{\sigma_x}\right)^2\right] dx \qquad -\infty < x < \infty. \quad (2.75)$$

If X follows a normal distribution with distribution parameters μ_x and σ_x, we have the following equations:

$$E(X) = \mu_x \qquad (2.76)$$

$$\sigma_x = E\left[(X - \mu_x)^2\right] = \sqrt{Var(X)} = \sigma_x. \qquad (2.77)$$

When two distribution parameters of a normal distribution μ_x and σ_x are known, the PDF of this normal distribution is fully specified, as shown in Equation (2.74). A normal distribution of a random variable with distribution parameters μ_x and σ_x can be expressed as

$$X = N(\mu_x, \sigma_x). \qquad (2.78)$$

The PDFs of a normal distribution with different distribution parameters are plotted in Figure 2.20. From Figure 2.20, the PDF of normal distribution has a bell-shaped curve with the symmetrical line $x = \mu_x$. The mean μ_x of normal distribution will control the horizontal location of the bell-shaped curve and the standard deviation σ_x will control the shape. When the means are the same, the bell-shaped curve will be thinner with a smaller standard deviation.

The CDF in Equation (2.75) of a normal distribution does not have an explicit theoretical solution. However, it can be easily calculated by using Excel and MATLAB.

In Microsoft Excel, the functions for calculating the PDF and the CDF of a normal distribution are:

$$f(x) = NORM.DIST(x, \mu_x, \sigma_x, false) \qquad (2.79)$$

$$F(x) = P(X \leq x) = NORM.DIST(x, \mu_x, \sigma_x, true). \qquad (2.80)$$

In MATLAB, the commands for calculating the PDF and CDF of a normal distribution are:

$$f(x) = normpdf(x, \mu_x, \sigma_x) \qquad (2.81)$$

$$F(x) = P(X \leq x) = normcdf(x, \mu_x, \sigma_x). \qquad (2.82)$$

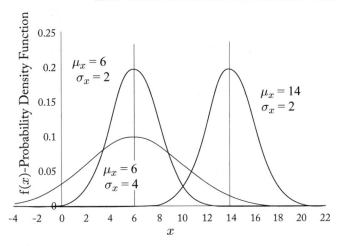

Figure 2.20: Several PDFs of normal distributions.

Example 2.39

If the tensile yield strength X of a ductile material follows a normal distribution with the mean $\mu_x = 61.2$ (ksi) and the standard deviation $\sigma_x = 4.25$ (ksi). Use Excel and MATLAB to calculate probability $P(50 \leq X \leq 70)$.

Solution:

By a Microsoft Excel function, that is, per Equation (2.79), we have:

$$P(50 \leq X \leq 70) = P(X \leq 70) - P(X \leq 50) = F(70) - F(50)$$
$$= NORM.DIST(70, 61.2, 4.25, true)$$
$$- NORM.DIST(50, 61.2, 4.25, true)$$
$$= 0.9808 - 0.0042 = 0.9766.$$

By MATLAB, that is, per Equation (2.81), we have:

$$P(50 \leq X \leq 70) = P(X \leq 70) - P(X \leq 50) = F(70) - F(50)$$
$$= normcdf(70, 61.2, 4.25) - normcdf(50, 61.2, 4.25)$$
$$= 0.9808 - 0.0042 = 0.9766.$$

∎

The normal distribution is one of the frequently-used distribution function. The probability theory proves that if $X_1, \ldots, X_n$ and are mutually independent normally distributed random variables, the linear function of these $X_1, \ldots, X_n$ and will also be a normal distributed random variable, that is,

$$Y = \sum_{i=1}^{n} a_i X_i \tag{2.83}$$

$$X_i = N(\mu_i, \sigma_i) \quad i = 1, \ldots, n \tag{2.84}$$

$$Y = N(\mu_y, \sigma_y) \tag{2.85}$$

$$\mu_Y = \sum_{i=1}^{n} (a_i \mu_i) \tag{2.86}$$

$$\sigma_Y = \sqrt{\sum_{i=1}^{n} (a_i \sigma_i)^2}, \tag{2.87}$$

where a_i is the constant coefficient. μ_i and σ_i are the mean and standard deviation of a normal distributed random variable X_i. μ_y and σ_y are the mean and standard deviation of a normal distributed random variable Y.

Example 2.40

The combined normal stress on a critical point of a component is the sum of normal stresses caused by mutually independent bending moment and axial loading. The normal stress σ_B caused by the bending moment follows a normal distribution with a mean $\mu_{\sigma_B} = 37.56$ (ksi) and a standard deviation $\sigma_{\sigma_B} = 4.56$ (ksi). The normal stress σ_A caused by the axial loading is also a normally distributed random variable with a mean $\mu_{\sigma_A} = 6.23$ (ksi) and a standard deviation $\sigma_{\sigma_A} = 1.02$ (ksi). Calculate the distribution parameters of the combined normal stress S_C and the probability of $P(S_C > 50)$.

Solution:

In this problem, the combined normal stress is the sum of the normal stresses caused by mutually-independent bending moment and axial loading, so we have:

$$S_C = \sigma_B + \sigma_A.$$

Per Equations (2.84), (2.85), and (2.86), we have:

$$\mu_{S_C} = \mu_{\sigma_B} + \mu_{\sigma_A} = 37.56 + 6.23 = 43.79 \text{ (ksi)}$$

$$\sigma_{S_C} = \sqrt{(\sigma_{\sigma_B})^2 + (\sigma_{\sigma_A})^2} = \sqrt{4.56^2 + 1.02^2} = 4.67 \text{ (ksi)}.$$

So, the combined normal stress, which is also a normal distributed random variable, can be expressed as:

$$S_C = N(\mu_{S_C}, \sigma_{S_C}) = N(43.79, 4.67) \text{ (ksi)}.$$

By Excel function per Equation (2.79), the probability of $P(S_C > 50)$ will be:

$$P(S_C > 50) = 1 - P(50 \leq S_C) = 1 - NORM.DIST(50, 43.79, 4.67, true)$$
$$= 1 - 0.9082 = 0.0918.$$

By MATLAB command per Equation (2.81), the probability of $P(S_C > 50)$ will be:

$$P(S_C > 50) = 1 - P(50 \leq S_C) = 1 - normcdf(50, 43.79, 4.67)$$
$$= 1 - 0.9082 = 0.0918.$$

■

Standard normal distribution: It is a special normal distribution with a mean $\mu = 0$ and standard deviation $\sigma = 1$. The PDF of the standard normal distribution, typically expressed by $\phi(z)$, as plotted in Figure 2.21, is:

$$f(z) = \phi(z) = \frac{1}{\sqrt{2\pi}} \exp\left[-\frac{1}{2}z^2\right] \qquad -\infty < z < \infty. \tag{2.88}$$

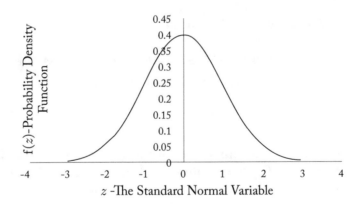

Figure 2.21: The plot of the PDF of a standard normal distribution.

The CDF of a standard normal distribution, typically expressed by $\Phi(z)$, is

$$F(z) = \Phi(z) = P(Z \leq z) = \int_{-\infty}^{x} \frac{1}{\sqrt{2\pi}} \exp\left[-\frac{1}{2}z^2\right] dz \qquad -\infty < z < \infty. \tag{2.89}$$

In Microsoft Excel, the functions for calculating the PDF and CDF of a standard normal distribution are:

$$\phi(z) = NORM.S.DIST(z, false) \tag{2.90}$$
$$\Phi(z) = P(Z \leq z) = NORM.S.DIST(z, true). \tag{2.91}$$

In MATLAB, the commands for calculating the PDF and CDF of a standard normally distributed variable are:

$$\phi(x) = normpdf(z) \tag{2.92}$$

$$\Phi(z) = P(Z \leq z) = normcdf(z). \tag{2.93}$$

Example 2.41

A random variable Z can be assumed to follow the standard normal distribution. Use Excel and MATLAB to calculate the probability of $P(-3 \leq Z \leq 3)$.

Solution:

By a Microsoft Excel function, that is, per Equation (2.91), we have:

$$P(-3 \leq Z \leq 3) = \Phi(3) - \Phi(-3)$$
$$= NORM.S.DIST(3, true) - NORM.S.DIST(-3, true)$$
$$= 0.9987 - 0.0013 = 0.9973.$$

By the MATLAB program, that is, per Equation (2.93), we have:

$$P(-3 \leq Z \leq 3) = \Phi(3) - \Phi(-3) = normcdf(3) - normcdf(-3)$$
$$= 0.9987 - 0.0013 = 0.9973.$$

∎

Since the PDF of a standard normal distribution is symmetrical about the vertical axis, that is, $z = \mu = 0$, as shown in Figure 2.22, the shaded area depicted in Figure 2.22 will be the same. The left shaded area is the probability $P(Z \leq -z)$, that is,

$$P(Z \leq -z) = \Phi(-z). \tag{2.94}$$

The right shaded area is the probability $P(Z > z)$, that is,

$$P(Z > z) = 1 - P(Z \leq z) = 1 - \Phi(z). \tag{2.95}$$

Since both shaded areas are the same as shown in Figure 2.22, we have the following equation per Equations (2.94) and (2.95):

$$\Phi(-z) = 1 - \Phi(z). \tag{2.96}$$

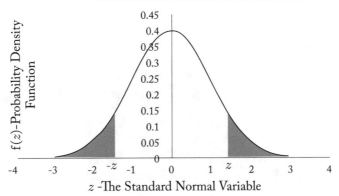

Figure 2.22: Schematic of $P(Z \leq z)$ and $P(Z \geq z)$.

Standard normal distribution table: The CDF of a standard normal variable Z, named as the standard normal distribution table, is shown in Table 2.10. This table can be used to determine the CDF $\Phi(z)$ of the standard normal variable for the value of $0 \leq z \leq 3$. For the value of $-3 \leq z \leq 0$, per Equation (2.96), we can also determine the CDF $\Phi(z)$ of a standard normal variable.

For a general normally distributed variable, we can use the following equation to convert it into the standard normal distributed variable. For $X = N(\mu_x, \sigma_x)$, Z will be the standard normal distributed variable if Z is defined by:

$$Z = \frac{X - \mu_x}{\sigma_x}. \tag{2.97}$$

Per Equation (2.97), we can have the following two equations to calculate the CDF or the probability of a normally distributed variable by using the standard normal distribution Table 2.10:

$$F(x) = P(X \leq x) = \Phi\left(\frac{x - \mu_x}{\sigma_x}\right) \tag{2.98}$$

$$P(a \leq X \leq b) = \Phi\left(\frac{b - \mu_x}{\sigma_x}\right) - \Phi\left(\frac{a - \mu_x}{\sigma_x}\right). \tag{2.99}$$

Example 2.42
If the tensile yield strength of a ductile material follows a normal distribution with a mean $\mu = 61.2$ (ksi) and standard deviation $\mu = 4.25$ (ksi). Use the standard normal distribution Table 2.10 to calculate probability $P(50 \leq X \leq 70)$.

Table 2.10: The standard normal distribution table

Standard Normal Distribution
$Z = N(0,1)$

$$\Phi(z) = P(Z \le z)$$
$$= \int_{-\infty}^{x} \frac{1}{\sqrt{2\pi}} \exp\left[-\frac{1}{2}z^2\right]dz$$

Z	0.00	0.01	0.02	0.03	0.04	0.05	0.06	0.07	0.08	0.09
0.0	0.5000	0.5040	0.5080	0.5120	0.5160	0.5199	0.5239	0.5279	0.5319	0.5359
0.1	0.5398	0.5438	0.5478	0.5517	0.5557	0.5596	0.5636	0.5675	0.5714	0.5753
0.2	0.5793	0.5832	0.5871	0.5910	0.5948	0.5987	0.6026	0.6064	0.6103	0.6141
0.3	0.6179	0.6217	0.6255	0.6293	0.6331	0.6368	0.6406	0.6443	0.6480	0.6517
0.4	0.6554	0.6591	0.6628	0.6664	0.6700	0.6736	0.6772	0.6808	0.6844	0.6879
0.5	0.6915	0.6950	0.6985	0.7019	0.7054	0.7088	0.7123	0.7157	0.7190	0.7224
0.6	0.7257	0.7291	0.7324	0.7357	0.7389	0.7422	0.7454	0.7486	0.7517	0.7549
0.7	0.7580	0.7611	0.7642	0.7673	0.7704	0.7734	0.7764	0.7794	0.7823	0.7852
0.8	0.7881	0.7910	0.7939	0.7967	0.7995	0.8023	0.8051	0.8078	0.8106	0.8133
0.9	0.8159	0.8186	0.8212	0.8238	0.8264	0.8289	0.8315	0.8340	0.8365	0.8389
1.0	0.8413	0.8438	0.8461	0.8485	0.8508	0.8531	0.8554	0.8577	0.8599	0.8621
1.1	0.8643	0.8665	0.8686	0.8708	0.8729	0.8749	0.8770	0.8790	0.8810	0.8830
1.2	0.8849	0.8869	0.8888	0.8907	0.8925	0.8944	0.8962	0.8980	0.8997	0.9015
1.3	0.9032	0.9049	0.9066	0.9082	0.9099	0.9115	0.9131	0.9147	0.9162	0.9177
1.4	0.9192	0.9207	0.9222	0.9236	0.9251	0.9265	0.9279	0.9292	0.9306	0.9319
1.5	0.9332	0.9345	0.9357	0.9370	0.9382	0.9394	0.9406	0.9418	0.9429	0.9441
1.6	0.9452	0.9463	0.9474	0.9484	0.9495	0.9505	0.9515	0.9525	0.9535	0.9545
1.7	0.9554	0.9564	0.9573	0.9582	0.9591	0.9599	0.9608	0.9616	0.9625	0.9633
1.8	0.9641	0.9649	0.9656	0.9664	0.9671	0.9678	0.9686	0.9693	0.9699	0.9706
1.9	0.9713	0.9719	0.9726	0.9732	0.9738	0.9744	0.9750	0.9756	0.9761	0.9767
2.0	0.9772	0.9778	0.9783	0.9788	0.9793	0.9798	0.9803	0.9808	0.9812	0.9817
2.1	0.9821	0.9826	0.9830	0.9834	0.9838	0.9842	0.9846	0.9850	0.9854	0.9857
2.2	0.9861	0.9864	0.9868	0.9871	0.9875	0.9878	0.9881	0.9884	0.9887	0.9890
2.3	0.9893	0.9896	0.9898	0.9901	0.9904	0.9906	0.9909	0.9911	0.9913	0.9916
2.4	0.9918	0.9920	0.9922	0.9925	0.9927	0.9929	0.9931	0.9932	0.9934	0.9936
2.5	0.9938	0.9940	0.9941	0.9943	0.9945	0.9946	0.9948	0.9949	0.9951	0.9952
2.6	0.9953	0.9955	0.9956	0.9957	0.9959	0.9960	0.9961	0.9962	0.9963	0.9964
2.7	0.9965	0.9966	0.9967	0.9968	0.9969	0.9970	0.9971	0.9972	0.9973	0.9974
2.8	0.9974	0.9975	0.9976	0.9977	0.9977	0.9978	0.9979	0.9979	0.9980	0.9981
2.9	0.9981	0.9982	0.9982	0.9983	0.9984	0.9984	0.9985	0.9985	0.9986	0.9986
3.0	0.9987	0.9987	0.9987	0.9988	0.9988	0.9989	0.9989	0.9989	0.9990	0.9990

Solution:
Per Equations (2.99) and (2.96), we have:

$$P\,(50 \leq X \leq 70) = \Phi\left(\frac{70 - 61.2}{4.25}\right) - \Phi\left(\frac{50 - 61.2}{4.25}\right)$$

$$= \Phi\,(2.07) - \Phi\,(-2.64) = \Phi\,(2.07) - [1 - \Phi\,(2.64)]$$

$$= \Phi\,(2.07) + \Phi\,(2.64) - 1.$$

From Table 2.10, we have

$$\Phi\,(2.07) = 0.9808, \qquad \Phi\,(2.64) = 0.9959.$$

Therefore, we have:

$$P\,(50 \leq X \leq 70) = \Phi\,(2.07) + \Phi\,(2.64)$$

$$= 0.9808 + 0.9959 - 1 = 0.9767.$$

■

2.12.5   LOG-NORMAL DISTRIBUTION

The lognormal distribution is a widely used type of distributions. Material ultimate strength, fatigue life, and fatigue strength might follow a lognormal distribution.

Lognormal distribution: A random variable X is a lognormal distribution if its logarithm $\ln X$ follows a normal distribution. The PDF of a lognormal distributed random variable X is:

$$f\,(x) = \frac{1}{\sqrt{2\pi}x\sigma_x}\exp\left[-\frac{1}{2}\left(\frac{x - \mu_x}{\sigma_x}\right)^2\right] \qquad x \geq 0, \tag{2.100}$$

where μ_x and σ_x are the mean and standard deviation of the lognormal distributed random variable X.

The CDF of a lognormal distributed random variable X is:

$$F\,(x) = P\,(X \leq x) = \int_0^x \frac{1}{\sqrt{2\pi}x\sigma_x}\exp\left[-\frac{1}{2}\left(\frac{x - \mu_x}{\sigma_x}\right)^2\right]dx \qquad x \geq 0. \tag{2.101}$$

The plots of lognormal distributed variables with the same mean $\mu_x = 0$ but different standard deviations $\sigma_x = 0.3, 0.6$, and 1 are shown in Figure 2.23. The PDF's shape of a lognormal distribution is not symmetrical anymore. With a bigger standard deviation, the shape of the PDF of a lognormal distribution is skewed toward the left.

For a lognormal distributed variable X, its logarithm $\ln X$ is a normal distribution. μ_x and σ_x, which are the mean and standard deviation of the lognormal distributed variable X, can

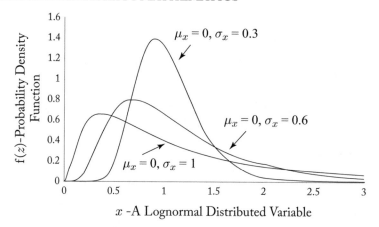

Figure 2.23: Plots of PDFs of lognormal distributions.

be directly calculated by the sample data of X. $\mu_{\ln x}$ and $\sigma_{\ln x}$, which are the mean and standard deviation of normal distributed variable $\ln X$, can be determined by using the sample data of $\ln X$ or calculated by using μ_x and σ_x.

If μ_x and σ_x are known, we can calculate the mean and standard deviation of the normally distributed variable $\ln X$:

$$\mu_{\ln x} = \ln\left(\frac{\mu_x^2}{\sqrt{\sigma_x^2 + \mu_x^2}}\right), \qquad \sigma_{\ln x} = \sqrt{\ln\left(1 + \frac{\sigma_x^2}{\mu_x^2}\right)}. \tag{2.102}$$

If $\mu_{\ln x}$ and $\sigma_{\ln x}$ are known, we can calculate the mean and standard deviation of the lognormally distributed variable X:

$$\mu_x = \exp\left(\mu_{\ln x} + \frac{\sigma_{\ln x}^2}{2}\right), \qquad \sigma_x = \sqrt{\left[\exp\left(\sigma_{\ln x}^2\right) - 1\right]\exp(2\mu_{\ln x} + \sigma_{\ln x}^2)}. \tag{2.103}$$

In Microsoft Excel, the functions for calculating the PDF and CDF of a lognormal distribution are:

$$f(x) = LOGNORM.DIST(x, \mu_{\ln x}, \sigma_{\ln x}, false) \tag{2.104}$$

$$F(x) = P(X \le x) = LOGNORM.DIST(x, \mu_{\ln x}, \sigma_{\ln x}, true). \tag{2.105}$$

In MATLAB, the commands for calculating the PDF and CDF of a lognormal distribution are:

$$f(x) = lognpdf(x, \mu_{\ln x}, \sigma_{\ln x}) \tag{2.106}$$

$$F(x) = P(X \le x) = logncdf(x, \mu_{\ln x}, \sigma_{\ln x}). \tag{2.107}$$

We can also use the standard normal distribution table to calculate the CDF of a lognormal distributed variable:

$$F(x) = P(X \leq x) = P(\ln X \leq \ln x) = \Phi \left(\frac{\ln x - \mu_{\ln x}}{\sigma_{\ln x}} \right). \tag{2.108}$$

Example 2.43
The life of brakes follow a lognormal distribution with a mean $\mu_x = 60001.4$ (miles) and a standard deviation $\sigma_x = 9100.6$ (miles), respectively. Use Excel, MATLAB, and the standard normal distribution table to calculate the probability $P(X > 40,000$ miles$)$.

Solution:
Per Equation (2.102), we have:

$$\mu_{\ln x} = \ln \left(\frac{\mu_x^2}{\sqrt{\sigma_x^2 + \mu_x^2}} \right) = \ln \left(\frac{60001.4^2}{\sqrt{9100.6^2 + 60001.4^2}} \right) = 10.9908$$

$$\sigma_{\ln x} = \sqrt{\ln \left(1 + \frac{\sigma_x^2}{\mu_x^2} \right)} = \sqrt{\ln \left(1 + \frac{9100.6^2}{60001.4^2} \right)} = 0.1508.$$

By using the Excel function per Equation (2.105), we have:

$$P(X > 40000) = 1 - P(X \leq 40000)$$
$$= 1 - LOGNORM.DIST(40000, 10.9908, 0.1508, true)$$
$$= 0.9955.$$

By using the MATLAB command per Equation (2.107), we have:

$$P(X > 40000) = 1 - P(X \leq 40000)$$
$$= 1 - logncdf(40000, 10.9908, 0.1508)$$
$$= 0.9955.$$

Per Equation (2.108), we have:

$$P(X > 40000) = 1 - P(X \leq 40000)$$
$$= 1 - \Phi \left(\frac{\ln x - \mu_{\ln x}}{\sigma_{\ln x}} \right) = 1 - \Phi \left(\frac{\ln 40000 - 10.9908}{0.1508} \right)$$
$$= 1 - \Phi(-2.6138).$$

By using the standard normal distribution table, that is, Table 2.10 and per Equation (2.96), we have:

$$P\left(X > 40000\right) = 1 - P\left(X \leq 40000\right)$$
$$= 1 - \Phi\left(-2.6138\right) = 1 - \left[1 - \Phi\left(2.6138\right)\right]$$
$$= \Phi\left(2.6138\right) \approx \Phi\left(2.61\right) = 0.9955.$$

∎

2.12.6 WEIBULL DISTRIBUTION

The Weibull distribution is one of the frequently used and versatile distributions. For example, it can be used to describe materials strength and fatigue life.

3-parameter Weibull distribution: The PDF of a three-parameter Weibull distribution is:

$$f\left(x\right) = \frac{\beta}{\eta} \left(\frac{x - \gamma}{\eta}\right)^{\beta - 1} e^{-\left(\frac{x - \gamma}{\eta}\right)^{\beta}} \qquad x \geq \gamma, \tag{2.109}$$

where η is a scale parameter, β is a shape parameter, and γ is a location parameter. All of these distribution parameters must be larger than zero.

The CDF of a three-parameter Weibull distribution is:

$$F\left(x\right) = 1 - e^{-\left(\frac{x - \gamma}{\eta}\right)^{\beta}} \qquad x \geq \gamma. \tag{2.110}$$

The graphs of three-parameter Weibull distribution with $\eta = 1$, $\beta = 1$, $\gamma = 0$, and 5 are shown in Figure 2.24.

From Figure 2.24, the two PDF graphs are identical, but the second one shifts to the right by the location parameter $\gamma = 5$. The three-parameter Weibull distribution can especially describe the distribution of material strength, which has a minimum proof strength. This minimum proof-strength would be equal to the location parameter. The most frequently used Weibull distribution is a two-parameter Weibull distribution.

2-parameter Weibull distribution: The PDF of two-parameter Weibull distribution, typically termed as Weibull distribution, is:

$$f\left(x\right) = \frac{\beta}{\eta} \left(\frac{x}{\eta}\right)^{\beta - 1} e^{-\left(\frac{x}{\eta}\right)^{\beta}} \qquad x \geq 0. \tag{2.111}$$

The CDF of a two-parameter Weibull distribution is:

$$F\left(x\right) = 1 - e^{-\left(\frac{x}{\eta}\right)^{\beta}} \qquad x \geq 0. \tag{2.112}$$

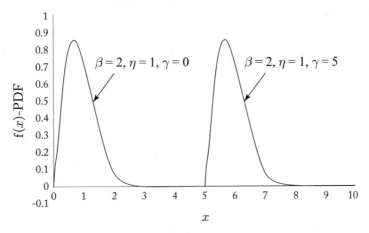

Figure 2.24: Graphs of three-parameter Weibull distribution.

Graphs of a Weibull distribution with different η and β are shown in Figure 2.25. With the same shape parameter $\beta = 2$, the PDF graph shifts to the right and become flattered when the scale parameters change from $\eta = 100$ to $\eta = 300$. With the same scale parameter $= 300$, the PDF graph shifts to the left and become flattered when the shape parameters change from $\beta = 3$ to $\beta = 2$. With the combination of different scale and shape parameters η and β, Weibull distribution can be used as the type of distribution of lots of different random variables.

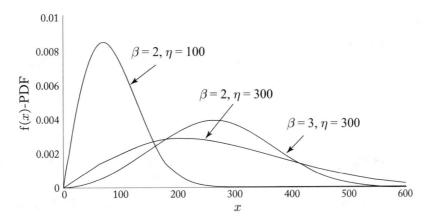

Figure 2.25: Graphs of Weibull distributions with different parameters.

The mean and standard deviation of a Weibull distribution are:

$$\mu_x = E\,(X) = \eta\Gamma\left(\frac{1}{\beta} + 1\right) \tag{2.113}$$

$$\sigma_x = Var\,(X) = \eta\sqrt{\Gamma\left(\frac{2}{\beta} + 1\right) - \left[\Gamma\left(\frac{1}{\beta} + 1\right)\right]^2}, \tag{2.114}$$

where $\Gamma\,(a)$ is the gamma function $\Gamma\,(a) = \int_0^\infty e^{-z}z^{a-1}dz$. The value of the gamma function can be determined by Microsoft Excel function or MATLAB command.

By Microsoft Excel,

$$\Gamma\,(a) = GAMMA(a). \tag{2.115}$$

By MATLAB command,

$$\Gamma\,(a) = gamma(a). \tag{2.116}$$

In Microsoft Excel, the functions for calculating the PDF and CDF of a Weibull distribution are:

$$f\,(x) = WEIBULL.DIST(x, \beta, \eta, false) \tag{2.117}$$
$$F\,(x) = P\,(X \le x) = WEIBULL.DIST(x, \beta, \eta, true). \tag{2.118}$$

In MATLAB, the commands for calculating the PDF and CDF of a Weibull distribution are:

$$f\,(x) = wblpdf(x, \eta, \beta) \tag{2.119}$$
$$F\,(x) = P\,(X \le x) = wblcdf(x, \eta, \beta). \tag{2.120}$$

Example 2.44

A bearing life X of a gearbox can be described by a two-parameter Weibull distribution with the scale parameter $\eta = 6000$ (hours) and the shape parameter $\beta = 0.62$. (1) Determine the mean and standard deviation of the bearing life. (2) Determine the probability $P\,(X > 3000)$.

Solution:

1. The mean and standard deviation.

   Per Equations (2.113) and (2.115) or (2.116), the mean of the bearing life will be:

$$\mu_x = \eta\Gamma\left(\frac{1}{\beta} + 1\right) = 6000 \times \Gamma\left(\frac{1}{0.62} + 1\right) = 6000 \times \Gamma\,(2.1629)$$

$$= 6000 \times 1.0802 = 6481.2 \ \text{(hours)}.$$

Per Equations (2.114) and (2.115) or (2.116), the standard deviation of the bearing life will be:

$$\sigma_x = \eta \sqrt{\Gamma\left(\frac{2}{\beta}+1\right) - \left[\Gamma\left(\frac{1}{\beta}+1\right)\right]^2}$$

$$= 6000 \times \sqrt{\Gamma\left(\frac{2}{0.62}+1\right) - \left[\Gamma\left(\frac{1}{0.62}+1\right)\right]^2}$$

$$= 6000 \times \sqrt{\Gamma(4.2258) - [\Gamma(2.1629)]^2} = 6000 \times \sqrt{8.0243 - 2.0840}$$

$$= 14623.64 \ (\text{hours}).$$

2. $P(X > 3000)$.

Per Equation (2.112), we can directly calculate the probability.

$$P(X > 3000) = 1 - P(X \leq 3000) = 1 - F(x) = 1 - \left[1 - e^{-\left(\frac{x}{\eta}\right)^{\beta}}\right]$$

$$= e^{-\left(\frac{3000}{6000}\right)^{0.62}} = 0.5217.$$

By using Microsoft Excel per Equation (2.118), we have:

$$P(X > 3000) = 1 - P(X \leq 3000) = 1 - WEIBULL.DIST\,(3000,0.62,6000, true)$$

$$= 1 - 0.4783 = 0.5217.$$

By using MATLAB per Equation (2.120), we have:

$$P(X > 3000) = 1 - P(X \leq 3000) = 1 - wbl\,(3000,6000,0.62)$$

$$= 1 - 0.4783 = 0.5217.$$

■

2.12.7 EXPONENTIAL DISTRIBUTION

The exponential distribution is one of the commonly used distributions. It is a simple distribution and is a special case of a Weibull distribution with the shape parameter $\beta = 1$. The exponential distribution is used to model the behavior of a device with a constant failure rate.

Exponential distribution: It is a one-parameter distribution. The PDF of an exponential distribution is:

$$f(x) = \lambda e^{-\lambda x} \qquad x \geq 0, \ \lambda > 0, \tag{2.121}$$

where λ is the parameter of the distribution, often called the rate parameter.

The CDF of an exponential distribution is:

$$F(x) = 1 - e^{-\lambda x} \qquad x \geq 0, \ \lambda > 0. \tag{2.122}$$

The graphs of the PDF of an exponential distribution with $\lambda = 0.5, 1$, and 1.5 are shown in Figure 2.26. The PDF of an exponential distribution has its maximum values at $x = 0$ and will be exponentially decreased.

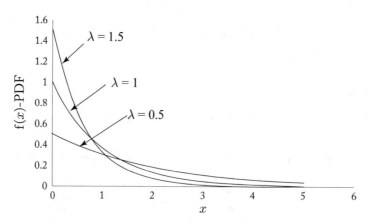

Figure 2.26: Graphs of the PDF of an exponential distribution with different λ.

The mean and the standard deviation of an exponential distribution are:

$$\mu_x = E(x) = \int_0^\infty x\lambda e^{-\lambda x} dx = \frac{1}{\lambda} \tag{2.123}$$

$$\sigma_x = \sqrt{Var(X)} = \sqrt{E[X^2] - \mu_x^2} = \frac{1}{\lambda}. \tag{2.124}$$

The PDF and CDF of an exponential distribution are easy to be calculated. Microsoft EXCEL and MATLAB still provide corresponding functions or commands for it.

In Microsoft Excel, the functions for calculating the PDF and CDF of an exponential distribution are:

$$f(x) = EXPON.DIST(x, \lambda, false) \tag{2.125}$$
$$F(x) = P(X \leq x) = EXPON.DIST(x, \lambda, true). \tag{2.126}$$

In the MATLAB, the commands for calculating the PDF and the CDF of an exponential distribution are:

$$f(x) = exppdf(x, 1/\lambda) \tag{2.127}$$
$$F(x) = P(X \leq x) = expcdf(x, 1/\lambda). \tag{2.128}$$

Example 2.45

The failure time X in the hour of a device is an exponential distribution with a failure rate $\lambda = 0.01 \left(\dfrac{1}{\text{hour}}\right)$. (1) Calculate the mean of the failure time. (2) Calculate probability $P(X \leq 25)$.

Solution:

1. Calculate the mean of the failure time.

   Per Equation (2.123), the mean of the failure time is:

   $$\mu_x = \frac{1}{\lambda} = \frac{1}{0.01} = 100 \text{ (hours)}.$$

2. $P(X \leq 25)$.

   Using the Microsoft Excel per Equation (2.126), the $P(X \leq 25)$ is:

   $$P(X \leq 25) = F(25) = EXPON.DIST(25, 0.01, true) = 0.2212.$$

   Using the MATLAB per Equation (2.128), the $P(X \leq 25)$ is:

   $$P(X \leq 25) = F(x) = expcdf(25, 1/0.01) = 0.2212.$$

   ∎

2.13 GOODNESS-OF-FIT TEST: χ^2 TEST

2.13.1 INTRODUCTION

When enough sample data (at least more than 30) is obtained, the histogram of the collected data can be drawn. Per the shape of the histogram, we can assume the type of distribution. For example, several histograms are displayed in Figure 2.27. From the histograms, the random variable with the histogram of Figure 2.27a might follow a normal distribution. The random variable with the histogram of Figure 2.27b might follow a lognormal distribution or a Weibull distribution. The random variable with the histogram of Figure 2.27c might follow an exponential distribution. The random variable with the histogram of Figure 2.27d might follow a uniform distribution.

The goodness-of-fit test is to determine whether a sample belongs to a hypothesized theoretical distribution. It estimates the adequacy of a fit by determining the difference between the frequency of occurrence of a random variable characterized by an observed sample and the expected frequency obtained from the hypothesized distribution.

There are different methods for conducting a goodness-of-fit test. One method is to use the χ^2 test with the Excel functions. Another method is to use the MATLAB command for the goodness-of-fit test, which is still based on the χ^2 test.

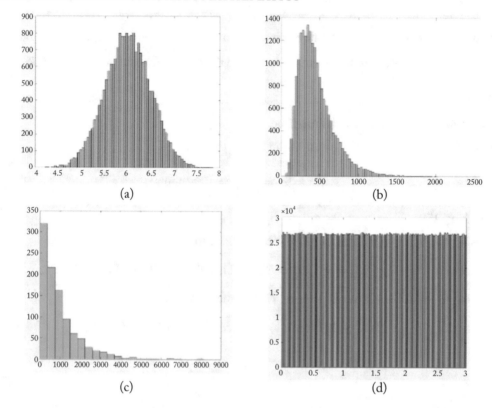

Figure 2.27: Several histograms.

2.13.2 THE CHI-SQUARE TEST

A Chi-square (χ^2) test can equally apply to any assumed distribution for a goodness-of-fit test. It is a commonly used and versatile test to check whether a set of data fits an assumed distribution.

The test statistic W for the Chi-square (χ^2) test is:

$$W = \chi^2 = \sum_{j=1}^{J} \frac{(o_j - e_j)^2}{e_j}, \tag{2.129}$$

where

- W—The test statistic which represents the difference between the sample data and the assumed distribution.

- W approximately follows a Chi-square (χ^2) distribution and will be described by the Chi-square (χ^2) distribution with $k = J - m - 1$ degrees of freedom

$$k = J - m - 1. \tag{2.130}$$

- k—The number of degrees of freedom of the Chi-square (χ^2) distribution for the test statistic W.

- J—The sample data X must be divided into J ($J \geq 6$) non-overlapping intervals. The lower limit for the first interval can be $-\infty$, and the upper limit of the last interval can be $+\infty$.

- m—The number of distribution parameters of the assumed distribution. For example, m is equal to 2 for a uniform distribution, a normal distribution, a lognormal distribution, a Weibull distribution, and 1 for an exponential distribution.

- o_j—The observed frequency of the sample data in the jth non-overlapping interval (x_{j-L}, x_{j-U}). o_j can be obtained through the histogram.

- e_j—The expected frequency in the jth non-overlapping interval (x_{j-L}, x_{j-U}) with the assumed distribution. e_j will be calculated from the assumed distribution by this formula:

$$e_j = N \times P\left(x_{j-L} \leq X \leq x_{j-U}\right) = N \times \left[F\left(x_{j-U}\right) - F\left(x_{j-L}\right)\right]. \tag{2.131}$$

- $F(x)$—The CDF of the assumed distribution whose distribution parameters are determined by the sample data.

- N—The total number of the sample data.

The hypothesis for the Chi-Square (χ^2) goodness-of-fit test is:

H_0: The random variable of sample data follows an assumed distribution whose distribution parameters are determined by the sample data.

With a confidence level $(1 - \alpha)$, the critical value $W_{critical}$ can be determined from the Chi-Square (χ^2) distribution:

$$W_{critical} = \chi^2_{1-a}(k), \tag{2.132}$$

$\chi^2_{1-a}(k)$ can be determined by the Chi-Square (χ^2) table, as shown in Table 2.11. In Microsoft Excel, $\chi^2_{1-a}(k)$ can be calculated by the following function:

$$\chi^2_{1-a}(k) = CHISQ.INV(1 - a, k). \tag{2.133}$$

In MATLAB, the $\chi^2_{1-a}(k)$ can be calculated by the following command:

$$\chi^2_{1-a}(k) = ch2inv(1 - a, k). \tag{2.134}$$

Two possible results for the Chi-Square (χ^2) goodness-of-fit test are as follows.

- If $W \geq W_{critical} = \chi^2_{1-a}(k)$, the hypothesis H_0 should be rejected with a confidence level $(1 - \alpha)$.

Table 2.11: The Chi-Square (χ^2) table

k	\multicolumn{6}{c}{α}					
	0.100	0.050	0.025	0.010	0.005	0.001
1	2.7055	3.8415	5.0239	6.6349	7.8794	10.8276
2	4.6052	5.9915	7.3778	9.2103	10.5966	13.8155
3	6.2514	7.8147	9.3484	11.3449	12.8382	16.2662
4	7.7794	9.4877	11.1433	13.2767	14.8603	18.4668
5	9.2364	11.0705	12.8325	15.0863	16.7496	20.5150
6	10.6446	12.5916	14.4494	16.8119	18.5476	22.4577
7	12.0170	14.0671	16.0128	18.4753	20.2777	24.3219
8	13.3616	15.5073	17.5345	20.0902	21.9550	26.1245
9	14.6837	16.9190	19.0228	21.6660	23.5894	27.8772
10	15.9872	18.3070	20.4832	23.2093	25.1882	29.5883
11	17.2750	19.6751	21.9200	24.7250	26.7568	31.2641
12	18.5493	21.0261	23.3367	26.2170	28.2995	32.9095
13	19.8119	22.3620	24.7356	27.6882	29.8195	34.5282
14	21.0641	23.6848	26.1189	29.1412	31.3193	36.1233
15	22.3071	24.9958	27.4884	30.5779	32.8013	37.6973
16	23.5418	26.2962	28.8454	31.9999	34.2672	39.2524
17	24.7690	27.5871	30.1910	33.4087	35.7185	40.7902
18	25.9894	28.8693	31.5264	34.8053	37.1565	42.3124
19	27.2036	30.1435	32.8523	36.1909	38.5823	43.8202
20	28.4120	31.4104	34.1696	37.5662	39.9968	45.3147
21	29.6151	32.6706	35.4789	38.9322	41.4011	46.7970
22	30.8133	33.9244	36.7807	40.2894	42.7957	48.2679
23	32.0069	35.1725	38.0756	41.6384	44.1813	49.7282
24	33.1962	36.4150	39.3641	42.9798	45.5585	51.1786
25	34.3816	37.6525	40.6465	44.3141	46.9279	52.6197
26	35.5632	38.8851	41.9232	45.6417	48.2899	54.0520
27	36.7412	40.1133	43.1945	46.9629	49.6449	55.4760
28	37.9159	41.3371	44.4608	48.2782	50.9934	56.8923
29	39.0875	42.5570	45.7223	49.5879	52.3356	58.3012
30	40.2560	43.7730	46.9792	50.8922	53.6720	59.7031
31	41.4217	44.9853	48.2319	52.1914	55.0027	61.0983
63	77.7454	82.5287	86.8296	92.0100	95.6493	103.4424
127	147.8048	154.3015	160.0858	166.9874	171.7961	181.9930
255	284.3359	293.2478	301.1250	310.4574	316.9194	330.5197
511	552.3739	564.6961	575.5298	588.2978	597.0978	615.5149
1023	1081.3794	1098.5208	1113.5334	1131.1587	1143.2653	1168.4972

- If $W < W_{critical} = \chi^2_{1-a}(k)$, the hypothesis $\boldsymbol{H_0}$ should not be rejected with a confidence level $(1 - \alpha)$.

α is a significant level, and $(1 - \alpha)$ is a confidence level. For the goodness-of-fit test, α should be at least 0.10 and typically is 0.05. The physical meaning of the confidence level $(1 - \alpha)$ is a probability of a judgment or statement. For example, when the hypothesis $\boldsymbol{H_0}$ is rejected with a confidence level $(1 - 0.05) = 0.95$, it means that this statement is true with a probability of 95%.

The procedure for the Chi-Square (χ^2) goodness-of-fit test can be summarized as follows.

- Step 1: Build the histogram of the sample data X. Based on the histogram or experience about the random variable, it can be hypothesized as a specific type of distribution, that is, the hypothesis H_0.

- Step 2: Calculate the mean μ_x and standard deviation σ_x of the sample data. Then use them to determine the distribution parameters of the hypothesized distribution.

- Step 3: Divide the sample data into J ($J \geq 6$) non-overlapping intervals.

- Step 4: o_j is obtained from the histogram. Use Equation (2.131) to calculate e_j in each non-overlapping interval.

- Step 5: Use Equation (2.129) to calculate the test statistic W.

- Step 6: Select a significant level a, typically 0.05.

- Step 7: Determine the critical value of $W_{critical} = \chi^2_{1-a}(k)$ by Equations (2.133) or (2.134) or the Chi-square Table 2.11.

- Step 8: If $W < W_{critical} = \chi^2_{1-a}(k)$, the hypothesis $\boldsymbol{H_0}$ should not be rejected with a confidence level $(1 - \alpha)$. Otherwise, the hypothesis $\boldsymbol{H_0}$ should be rejected.

Example 2.46

One hundred sample data of a material's yield strength is listed in Table 2.12. Make proper assumption of the type of distribution for this set of data and then use the Chi-square (χ^2) goodness-of-fit test to check whether it follows a normal distribution.

Solution:

- Step 1: Histogram of the sample data X.

    The histogram of this set of 100 sample data is shown in Figure 2.28. Per this histogram, it might be a normal distribution. So, we can make this assumption:

    $\boldsymbol{H_0}$: The material's yield strength follows a normal distribution.

Table 2.12: Sample data of material's yield strength

Yield Strength Data (ksi)
63.0, 56.4, 57.7, 56.4, 62.5, 55.9, 54.5, 62.4, 57.6, 61.5, 58.0, 61.0, 47.1, 62.0, 68.8, 64.8, 64.2, 58.5, 60.6, 60.9,
60.2, 55.6, 58.0, 58.8, 57.3, 63.8, 72.1, 53.2, 60.3, 52.6, 66.1, 66.6, 51.6, 69.2, 54.4, 60.8, 65.1, 60.4, 70.9, 73.2,
60.4, 50.4, 57.9, 55.5, 56.0, 54.2, 60.1, 70.0, 56.2, 58.3, 65.4, 65.6, 62.1, 64.7, 64.0, 57.8, 61.8, 58.2, 55.3, 62.8,
60.0, 58.8, 61.1, 64.5, 61.6, 60.7, 61.1, 60.0, 55.9, 63.6, 66.6, 57.1, 69.1, 58.8, 58.5, 55.3, 55.8, 55.0, 61.4, 67.5,
62.3, 63.9, 66.3, 47.5, 58.9, 61.4, 63.2, 53.3, 54.8, 63.6, 59.1, 62.4, 55.0, 51.7, 63.4, 65.5, 67.7, 52.2, 53.1, 52.5

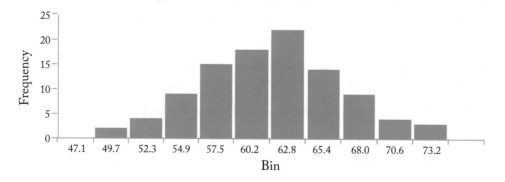

Figure 2.28: Histogram of the sample data.

- Step 2: The mean μ_x and the standard deviation σ_x of the sample data.

  Copy the sample data into a Microsoft Excel file. The mean and standard deviation based on these sample data are:

  $$\mu_x = 60.1 \ (\text{ksi}), \qquad \sigma_x = 5.23 \ (\text{ksi}).$$

  So, the hypothesized normal distribution will have the distribution parameters: the mean $\mu_x = 60.1$ (ksi) and the standard deviation $\sigma_x = 5.23$ (ksi).

- Step 3: Divide the sample data into *ten* non-overlapping intervals.

  The ten non-overlapping intervals are shown in the first two columns of Table 2.13.

- Step 4: o_j and e_j in each non-overlapping interval.

  The o_j can be directly obtained from the histogram. e_j can be calculated per Equation (2.131). For example, e_2 in the second non-overlapping interval (49.7, 52.36) is cal-

Table 2.13: The calculation results for the goodness-of-fit test

Non-Overlapping Interval		o_j	e_j	$\dfrac{(o_j - e_j)^2}{e_j}$
Lower range x_{j-L}	Upper range x_{j-U}			
$-\infty$	49.7	2	2.338724	0.049058
49.7	52.3	4	4.481793	0.051793
52.3	54.9	9	9.282114	0.008574
54.9	57.5	15	15.06146	0.000251
57.5	60.2	18	19.14866	0.068903
60.2	62.8	22	19.07542	0.448386
62.8	65.4	14	14.88931	0.053116
65.4	68.0	9	9.105954	0.001233
68.0	70.6	4	4.363151	0.030226
70.6	$+\infty$	3	2.253421	0.247348
		$W = \displaystyle\sum_{j=1}^{10} \dfrac{(o_j - e_j)^2}{e_j}$		0.958889

culated with the function in Microsoft Excel as:

$$e_2 = N \times P\,(x_{2-L} \leq X \leq x_{2-U}) = N \times [F\,(x_{2-U}) - F\,(x_{2-L})]$$
$$= 100 \times [NORM.DIST\,(52.3, 60.1, 5.23, TRUE)$$
$$- NORM.DIST\,(49.7, 60.1, 5.23, TRUE)]$$
$$= 4.481793.$$

- Step 5: The test statistic W.

  The test statistic W can be calculated per Equation (2.129). In the above table, the last column is the values of $\frac{(o_j - e_j)^2}{e_j}$ in each non-overlapping interval. According to the above table, we have the test statistic
  $$W = 0.959.$$

- Step 6: Select a significant level $a = 0.05$.

- Step 7: $W_{critical} = \chi^2_{1-a}(k)$.

  By Equations (2.132) or (2.133) or the Chi-square Table 2.11:

  $$W_{critical} = \chi^2_{1-a}\,(k) = \chi^2_{1-0.05}\,(10 - 2 - 1) = \chi^2_{0.95}\,(7) = 14.07.$$

- Step 8: The Chi-Square goodness-of-fit test.

Since $W = 0.989 < \chi^2_{0.95} (7) = 14.07$, the hypothesis $\boldsymbol{H_0}$ should not be rejected with a confidence level 95%, that is, the material's yield strength follows the normal distribution with a mean $\mu_x = 60.1$ (ksi) and a standard deviation $\sigma_x = 5.23$ (ksi). ∎

2.13.3   THE CHI-SQUARE (χ^2) GOODNESS-of-FIT TEST BY THE MATLAB PROGRAM

The MATLAB can be used to conduct the Chi-square (χ^2) goodness-of-fit test. Suppose that the collected sample data X is arranged in one column matrix and stored as a Microsoft Excel file. Import the sample data X into a MATLAB program and then use the command "histogram (X)" to create its histogram. Per the histogram or previous experience about the sample data, we can assume that the sample data follow a specific distribution.

In MATLAB software, there are several commands related to the goodness-of-fit test. For the goodness-of-fit test, two commands in the MATLAB program are needed. The first command $fitdist(X, distname)$ is to create the assumed distribution with the distribution parameters which are determined by the sample data X

$$pd = fitdist(X, distname), \qquad (2.135)$$

where

X—is the sample data and is expressed as a one-column matrix.

$distname$—is the name of the assumed distribution. "wbl" for a Weibull distribution and "exp" for an exponential distribution. A full list of $distname$ can be found through the "Help" in MATLAB software.

pd—is just any variable name for representing the assumed distribution.

The second command $chi2gof(X,'CDF', pd,'Alpha', value)$ is to conduct the Chi-square (χ^2) goodness-of-fit test.

$$h = chi2gof(X,'CDF', pd,'Alpha', value) \qquad (2.136)$$
$$h = chi2gof(X,'CDF', pd), \qquad (2.137)$$

where

X—is the same as that in Equation (2.135).

$'CDF'$—will not be changed and means that the comparison between the sample data and the assumed distribution will be through the CDF.

pd—should be the same variable name as that in Equation (2.135).

'Alpha', *value*—is the significant level *a* and its value. The value of the significant level *a* can be 0.10, 0.05, 0.01, and so on. However, the significant level *a* must be at least 0.10. If the significant level *a* has a default value, that is, 0.05, the command in Equation (2.137) can be used.

h—will be equal to 0 or 1. If *h* is equal to 0, the hypothesis should not be rejected; that is, the sample data can be properly described by the assumed distribution. If *h* is equal to 1, the hypothesis should be rejected; that is, the sample data cannot be properly described by the assumed distribution.

Example 2.47
A set of 100 sample data of material strength is listed in Table 2.14 and is stored in an Excel file with the name "Example2-44.xls." (1) Display its histogram. (2) Use the MATLAB program to check whether it follows a lognormal distribution with a significant level $a = 0.05$. (3) Use the MATLAB program to check whether it follows a Weibull distribution with a significant level $a = 0.05$. (4) Use the MATLAB program to check whether it follows a normal distribution with a significant level $a = 0.05$.

Table 2.14: One hundred sample data of material strengh (ksi)

12.47, 12.15, 22.01, 16.66, 10.57, 28.11, 27.98, 47.46, 60.95, 17.63, 37.9, 5.91, 9.69, 6.1, 13.82, 39.25, 44.65, 42.83, 85.33, 14.06, 8.13, 2.66, 23.99, 27.36, 23.28, 13.52, 63.75, 8.83, 158.65, 21.83, 16.04, 46.82, 20.22, 55.95, 74.14, 7.1, 37.27, 24.6, 48.51, 4.96, 49.78, 27.69, 77.36, 3.79, 19.79, 18.63, 23.9, 153.06, 27.72, 74.08, 9.89, 6.99, 14.76, 10.07, 14.25, 30.84, 19.63, 9.34, 61.22, 10.37, 54.44, 32.57, 6.88, 41.54, 49.41, 30.09, 16.82, 29.09, 72.4, 36.42, 25.31, 9.25, 11.33, 10.06, 18.42, 23.85, 29.87, 38.07, 24.71, 56.74, 25.72, 3.92, 5.53, 37.66, 45.59, 17.19, 79.8, 6.51, 13.17, 15.55, 12.28, 13.91, 10.79, 38.44, 19.56, 3.86, 31.36, 27.29, 71.44, 24.31,

Solution:

1. Display its histogram.

   The MATLAB program for display the histogram with a 10 bin is:

   ```
   %Import the data from an Excel file
   X=xlsread('Example 2.44'),
   % Create a 10-bin histogram
   histogram(X,10)
   ```

The histogram of this sample data is shown in Figure 2.29. From the histogram, the sample data might follow a lognormal or Weibull distribution. Since the histogram is significantly skewed toward the left, it could not be a normal distribution.

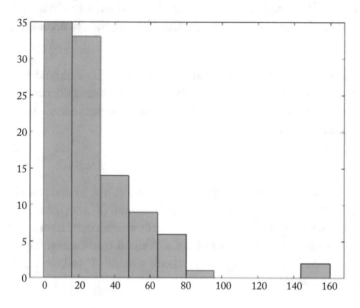

Figure 2.29: The histogram of sample data.

2. A lognormal distribution.

The hypothesis H_0: The material strength follows a lognormal distribution.

The MATLAB program for this goodness-of-fit test is:

```
% The goodness-of-fit test on the data in the Excel
% file "Example2.43.xls"
X=xlsread('Example2.44');
Q2=fitdist(X,'logn')
h=chi2gof(X,'CDF',Q2)
```

After running the above MATLAB program, the result of "h" is 0. The hypothesis should not be rejected. So, the material strength can be properly described by a lognormal distribution.

3. A Weibull distribution.

The hypothesis H_0: the material strength follows a Weibull distribution.

The MATLAB program for this goodness-of-fit test is:

```
% The goodness-of-fit test on the data in the Excel
% file "Example2.43.xls"
X=xlsread('Example2.44');
Q3=fitdist(X,'wbl')
h=chi2gof(X,'CDF',Q3)
```

After running the above MATLAB program, the result for "h" is 0. The hypothesis should not be rejected. So, the material strength can also be properly described by a Weibull distribution.

4. A normal distribution.

The hypothesis H_0: the material strength follows a normal distribution.

The MATLAB program for this goodness-of-fit test is:

```
% the goodness-of-fit test on the data in the Excel
% file "Example2.43.xls"
X=xlsread('Example2.44');
Q4=fitdist(X,'norm')
h=chi2gof(X,'CDF',Q4)
```

After running the above MATLAB program, the result of "h" is 1. The hypothesis should be rejected. So, the material strength cannot be properly described by a normal distribution. ■

2.14   REFERENCES

[1] Rao, S. S., *Reliability Engineering*, Pearson, 2015.

[2] Haugen, E. B., *Probability Approach to Design*, Wiley, New York, 1968.

[3] Schwarzlander, H., *Probability Concepts and Theory for Engineers*, Wiley, 2011. DOI: 10.1002/9781119990895.

[4] Miller, I. and Freund, J. E., *Probability and Statistics for Engineers*, 3rd ed., Prentice Hall, 1985.

[5] Modarres, M., Kaminskiy, M., and Kirvtsov, V., *Reliability Engineering and Risk Analysis: A Practical Guide*, 2nd ed., CRC Press, 2010. DOI: 10.1201/9781315382425.

2.15   EXERCISES

2.1.   What is an experiment? What is an event? Use one example to explain them.

2.2.   What is the sample space? What is the sample point? Use one example to explain them.

2.3.   For an experiment of throwing a dice, the two sets are $A = \{1, 2, 3\}$, $B = \{1, 3, 4\}$. Determine $A \cup B$ and $A \cap B$.

2.4.   Toss a coin for a total of 1000 times. The occurrence of "head" was 489. What is the probability of occurrence of the "head"? What is the probability of occurrence of the "tail"?

2.5.   When two dice are rolled, determine the probability of realizing the sum of two numbers less than six.

2.6.   A box contains four white balls and five green balls. A ball is selected randomly from the box, and its color is noted. The ball is then returned to the box, and another ball is selected randomly, and its color is noted. Determine the probabilities of realizing the following events:

   (a) two red balls, and

   (b) first ball red and the second ball green.

2.7.   Describe and explain a random event and a random variable. List two examples for each in mechanical designs.

2.8.   For an experiment of rolling a dice, $\Omega = \{1, 2, 3, 4, 5, 6\}$, $A = \{1, 2, 3\}$, $B = \{3, 4, 5, 6\}$, $C = \{5, 6\}$. Determine the probabilities: $P(A \cup B)$, $P(A \cap B)$, $P(A \cup C)$, $P(A \cap C)$.

2.9.   An experiment consists of tossing a coin four times.

   (a) Find the PMF and distribution function for the number of heads realized.

   (b) Find the probability of realizing heads at least two times.

2.10.   A small village has two grocery stores A and B. It is estimated that 40% of potential customers do business only with the store A, 30% only with store B, and 20% with both stores A and B. The remaining 10% of customers do business with none of the stores A and B. If E_1 (E_2) denote the event of a randomly selected customer doing business with the store A (B), calculate the probabilities: $P(E_1 \cup E_2)$, $P(E_1 \cap E_2)$, $P(\overline{E_1})$, and $P(\overline{E_1} \cup E_2)$.

2.11.   A batch of 195 specimens of composite material has the data shown in Table 2.15 when they are tested for strength and density.

Table 2.15: Strength and density

	High Density	Low Density
High Strength	142	22
Low Strength	16	15

(a) For a specimen randomly selected from the batch, calculate the probability that both its strength and density are low.

(b) For a specimen randomly selected from the batch, calculate the probability that its strength is low, but density is high.

(c) Let E_1 denotes that a specimen has high strength and E_2 denotes that a specimen has a high density. Are the events E_1 and E_2 independent to each other?

2.12. The inspection of a batch 1000 castings produced in an engine production plant for the presence of defect yielded the data in Table 2.16.

Table 2.16: Presence of defect

Type of defect	Location of Defect: Inside	Location of Defect: On the surface	Total
Blowholes	600	230	830
Crack	75	95	170
Total	675	325	1000

For a casting selected randomly from the batch, determine the following.

(a) The probability that the casting has a crack.

(b) The probability that the casting has a defect on the surface.

(c) The probability that the case has a blowhole which is on the surface.

(d) The probability that the casting has a crack which is inside.

2.13. An applied mathematics class consists of 35 male and 15 female students. Among the male students, 15 are science majors, and 20 are engineering majors. Among the female students, 5 are science majors and 10 are engineering majors. E_1 denotes a female student and E_2 a male student. M_1 denotes a student in a science major and M_2 a student in an engineering major.

(a) Calculate $P(E_1 | M_1)$.

      (b) Calculate $P(M_2 \,|\, E_2)$.

      (c) Calculate $P(E_1 M_1)$.

      (d) Calculate $P(E_2 M_1)$.

2.14. In a thermal power plant, four components—the feed water pump, boiler, turbine, and generator—are connected one after another as shown in the following figure. The probabilities of each component are also listed in Figure 2.30. It is assumed that the failure of any component is independent of each other. Calculate the probability that all four components are functional.

Figure 2.30: Schematic of the block diagram for Problem 2.14.

2.15. In an engine manufacturing plant, 70% of the crankshafts are ground by a machinist R and 30% by another machinist S. It is known from previous experience that the crankshafts ground by the machinists R and S contain 2% and 3% defective units, respectively. What is the probability of defective units?

2.16. A batch of 1000 piston O-ring manufactured in a factory contains 35 defective ones. Two piston rings are randomly selected from the batch, one at a time, without replacement. E_1 denotes a defective O-ring in the first selection and E_2 a defective O-ring in the second selection.

      (a) Calculate the probability $P(E_1)$.

      (b) Calculate the probability $P(E_2)$.

      (c) Calculate the probability $P(E_1 E_2)$.

2.17. The percentage of mechanical, civil, biomedical engineering students in an Engineering Statics course is 37, 40, and 23, respectively. The probability of a mechanical, civil, biomedical engineering student getting the grade A is 0.22, 0.15, and 0.32, respectively.

      (a) Calculate the probability of a student getting grade A in this class.

      (b) If a student is randomly selected and has a grade A, calculate the probability of the student belonging to mechanical engineering.

2.18. Describe and explain mean, standard deviation, and coefficient of variance.

2.19. The diameter X of manufactured shafts (unit in inch) by two operators are listed in Table 2.17.

Table 2.17: Manufactured shafts

Operator A	2.99, 3.21, 3.33, 3.31, 3.38, 3.29, 3.25, 3.08
Operator B	3.01, 2.89, 3.45, 3.89, 2.76, 3.34, 3.01, 3.49

(a) Calculate the mean and standard deviation of the diameter by each operator.

(b) Which of the operators is better based on the data? Why?

2.20. Tensile test data of 100 specimens of a material are listed in Table 2.18.

Table 2.18: Tensile test data

Yield Strength Data
55.8, 50.3, 59.4, 56.4, 55.6, 57.4, 56.2, 50.7, 54.3, 54.3, 52.6, 62.7, 51.0, 52.8, 51.7, 49.6, 54.1, 53.7, 62.0, 53.9,
50.1, 62.4, 60.7, 53.9, 48.1, 53.0, 54.3, 56.3, 53.8, 57.0, 56.8, 49.2, 50.6, 51.6, 52.7, 53.5, 55.1, 41.1, 52.9, 60.7,
50.1, 59.3, 56.6, 54.9, 55.8, 47.8, 54.6, 62.4, 55.5, 55.2, 51.6, 54.9, 56.1, 57.0, 53.3, 53.9, 64.3, 44.6, 65.3, 56.6,
59.6, 47.3, 52.3, 53.7, 56.9, 47.3, 57.2, 49.4, 55.3, 58.0, 56.5, 60.0, 59.6, 52.0, 56.2, 50.7, 48.9, 59.3, 55.0, 54.7,
59.2, 57.7, 56.6, 60.8, 59.3, 56.1, 51.8, 52.0, 60.5, 47.6, 54.9, 46.0, 59.7, 59.0, 55.0, 54.7, 43.6, 57.7, 44.9, 44.3

(a) Use six equal non-overlapping bin intervals and manually count the number of test data in each bin interval and draw the histogram.

(b) Input the data into one column in an Excel file and then use the histogram tool in Excel to draw the histogram with ten equal bin intervals.

(c) Compile a MATLAB program to read the data through the Excel file and use the histogram command to draw the histogram with 11 bins.

2.21. One hundred sampling data of a random variable are listed in Table 2.19.

Table 2.19: Random variable

The Sampling Data
41.2, 25.7, 34.9, 60.2, 35.2, 19.7, 21.6, 30.4, 30.1, 21.5, 36.2, 62.4, 29.2, 29.9, 11.0, 22.5, 16.5, 27.3, 43.1, 70.9,
81.4, 31.2, 28.2, 49.8, 42.3, 48.6, 48.9, 15.8, 43.4, 31.6, 22.6, 23.4, 62.8, 22.1, 17.8, 36.9, 90.5, 33.5, 38.6, 20.7,
76.5, 35.3, 43.2, 20.6, 50.8, 40.2, 18.6, 33.8, 26.4, 35.0, 29.2, 30.1, 19.8, 28.2, 48.6, 79.2, 18.5, 71.0 ,36.0, 28.5,
23.4, 50.2, 23.4, 26.3, 51.5, 41.2, 51.9, 42.6, 27.1, 25.6, 49.3, 23.7, 59.9, 49.2, 38.2, 33.2, 39.8, 19.1, 31.3, 15.2,
86.3, 108.7, 44.9, 24.0, 122.6, 43.6, 38.4, 22.4, 19.4, 13.7, 26.8, 19.6, 45.8, 28.3, 80.2, 70.5, 35.9, 28.7, 58.9, 18.7,

(a) Use ten equal non-overlapping bin intervals and manually count the number of test data in each bin interval and draw the histogram.

(b) Input the data into one column in an Excel file and then use the histogram tool in Excel to draw the histogram with ten equal bin intervals.

(c) Create a MATLAB program to read the data through the Excel file and use the histogram command to draw the histogram with eight bins.

2.22. A random variable X has the following PDF:

$$f(x) = \begin{cases} 0 & x < 40 \\ 0.025 & 40 \leq x \leq 80 \\ 0 & x > 80. \end{cases}$$

Calculate its mean, standard deviation, and coefficient of variance.

2.23. The applied force X on a component can be treated as a random variable and described by following PDF,

$$= \begin{cases} \dfrac{x}{24} & 0 \leq x \leq 6 \\ \dfrac{8-x}{8} & 6 \leq x \leq 8. \end{cases}$$

(a) Determine the mean, standard devaluation of the random variable X.

(b) Determine the CDF.

(c) Calculate the probability of $P(X > 1.5)$.

2.24. The life of a device X (in thousands of hours) is represented by the PDF:

$$f(x) = 2.15e^{-2.15x} \qquad x \geq 0.$$

(a) Determine the mean, standard devaluation of the random variable X.

(b) Determine the CDF.

(c) Calculate the probability of $P(X > 2.5)$.

2.25. A random variable is a realizing number of rolling a dice.

(a) Determine its PMF.

(b) Determine its CDF.

(c) Calculate the mean, standard deviation, and coefficient of variance of this random variable.

Table 2.20: Emails per minute

X	0	1	2	3	4	5	6	7	8
$p(x)$	0.005	0.06	0.22	0.31	0.32	0.06	0.015	0.005	0.005

2.26. The number of emails X that is received by an admission per minute has the following PMF $p(x)$ (Table 2.20).

    (a) Calculate the mean, standard deviation, and coefficient of the variance.

    (b) Determine its CDF.

2.27. A town has three sets of water pump systems. It is assumed that the probability of a water-pump system functioning without failure for one year is 0.95. It is also assumed that the performance of these three water pump systems can be described by a binomial distribution.

    (a) Determine its PMF.

    (b) Determine its CDF.

    (c) Calculate its mean, standard deviation.

    (d) Calculate the probability of the event, which is at least one water-pump system functioning without a failure.

2.28. It is assumed that 65% of a household with an installation of a solar heat system will have a utility bill in reduction by at least one-third.

    (a) What is the probability that a utility bill will be reduced by at least one third in four out of five households with the installation of solar heat systems?

    (b) What is the probability that a utility bill will be reduced by at least one third in at least three out of five households with the installation of solar heat systems?

2.29. If the probability is 0.25 that any person will dislike the taste of a new ice cream, what is the probability that 6 out of 18 randomly selected persons will dislike it?

2.30. The arrival of cars at a UPS facility is a Poisson process with a mean arrival rate of three cars per hour.

    (a) Determine the probability of the event that there are five arrival cars in two hours.

    (b) Determine the probability of the event that there are at least two arrival cars in two hours.

2.31. The number of defects in a continuous welded joint in an assembly is 0.05 per meter length. If the number of defects in the weld follows a Poisson distribution. Find the following.

(a) The probability of finding exactly one defect in a 10-m length of the weld.

(b) The probability of finding at most three defects in a 10-m length of the weld.

(c) The probability of finding at least one defect in a 2-m length of the weld.

2.32. The number of emails received of an admission account can be described as a Poisson process with a mean value of 6 per hour. Determine the following.

(a) The probability that no email is received in 30 min.

(b) The probability that at least 2 emails is received in 30 min.

2.33. The hardness of a component follows the uniform distribution with the following PDF:

$$f_H(h) = \begin{cases} 0 & h < 245 \ (HB) \\ \dfrac{1}{50} & 245 \leq h \leq 295 \ (HB) \\ 0 & h > 295 \ (HB). \end{cases}$$

(a) Calculate the mean and standard deviation.

(b) Calculate the probabilities: $P(100 < H < 260)$, $P(H < 300)$, $P(250 < H < 280)$.

2.34. The pressure X (psi) on a device is treated as a random variable and can be described by a uniform distribution. The PDF is:

$$f(x) = \begin{cases} \dfrac{1}{4000} & 4500 \leq x \leq 8500 \\ 0 & \text{otherwise.} \end{cases}$$

(a) Determine the mean and standard deviation.

(b) Determine the CDF.

(c) Calculate the following probability: $P(X \leq 3000)$, $P(2000 \leq X \leq 7500)$, and $P(6000 \leq X)$.

2.35. Loading X in lb on a component is a random variable and can be described by a uniform distribution:

$$f(x) = \begin{cases} \dfrac{1}{200} & 50 \leq x \leq 250 \\ 0 & \text{otherwise.} \end{cases}$$

(a) Determine the mean and standard deviation.

(b) Determine the CDF.

(c) Calculate the following probability: $P(X \leq 220)$, $P(0 \leq X \leq 250)$, and $P(200 \leq X)$.

2.36. The normal distribution is widely used in reliability engineering. Use MATLAB to plot PDFs of three normal distributions: $(\mu_1 = 100, \sigma_1 = 50)$; $(\mu_2 = 100, \sigma_2 = 5)$; $(\mu_3 = 50, \sigma_3 = 5)$.

2.37. For a normal distribution random variable X with $\mu_x = 100$ and $\sigma_x = 20$, calculate $P(0 < X < 150)$; $P(50 < X < 200)$ and $P(X < 120)$.

2.38. It is assumed that the fatigue life T (cycles) of a component follows a log-normal distribution. The mean μ_T and the standard deviation σ_T based on a set of test data are: $\mu_T = 45000$ (cycles) and $\sigma_T = 3500$ (cycles).

(a) Determine parameters $\mu_{\ln T}$ and $\sigma_{\ln T}$ of the lognormal distribution.

(b) Use Excel to plot the PDF of this lognormal distribution.

(c) Calculate following probabilities: $P(T > 30{,}000)$, $P(20{,}000 < T < 40{,}000)$, $P(T < 60{,}000)$.

2.39. Show the PDF and CDF of two-parameters of Weibull distribution.

(a) Use MATLAB to plot the PDFs of Weibull PDFs with the parameters (a) $\eta = 100, \beta = 1.25$; (b) $\eta = 100, \beta = 4$; and (c) $\eta = 10, \beta = 1.25$.

(b) For the Weibull distribution with the distribution parameter $\eta = 20, \beta = 1.5$, calculate its mean and standard deviation.

(c) For the Weibull distribution with the distribution parameter $\eta = 20, \beta = 1.5$, calculate the probabilities: $P(X < 80)$, $P(X < 40)$, $P(10 < X < 90)$.

2.40. A normally distributed random variable X has a mean 100 and a standard deviation 20.

(a) Calculate the probability $P(50 < X < 150)$.

(b) Recalculate the probability $P(50 < X < 150)$ if the mean is kept the same, but the standard deviation is reduced to 10.

2.41. The fatigue life X (cycles) of the component at a given stress level is a random variable and can be described by a log-normal distribution. The mean and standard devotion of the test data are $\mu_X = 55{,}000$ (cycles) and $\sigma_X = 6000$ (cycles).

(a) Determine the distribution parameters: $\mu_{\ln X}$ and $\sigma_{\ln X}$.

(b) Use the Excel function to calculate the probability: $P(X \geq 40{,}000)$.

(c) Use the MATLAB command to calculate the probability: $P(X \leq 20,000)$.

2.42. The measured resistance R of a type of resistor manufactured by a company can be described by a normal distribution with a mean $\mu_R = 120.0(\Omega)$ and a standard deviation $\sigma_R = 0.6(\Omega)$.

    (a) Use the Excel function to calculate the probability $P(118.5 \leq R \leq 121.5)$. Explain the physical meaning of the calculated value.

    (b) If there are 200 pieces of this resistor, estimate the number of these resistors which has the resistance in the range $(118.5 \leq R \leq 121.5)$.

2.43. Use MATLAB to plot the PDFs and the CDFs of a normal distribution with three different sets of distribution parameters: (1) $\mu = 100$ and $\sigma = 5$; (2) $\mu = 100$ and $\sigma = 10$; and (3) $\mu = 100$ and $\sigma = 30$. Discuss the effect of standard deviation σ on a normal PDF.

2.44. Use Excel to plot the PDFs and CDFs of a normal distribution with three different sets of distribution parameters: (1) $\mu = 10$ and $\sigma = 10$; (2) $\mu = 50$ and $\sigma = 10$; and (3) $\mu = 100$ and $\sigma = 10$. Discuss the effect of the mean on a normal PDF.

2.45. The diameter X of a forged crankshaft follows a normal distribution. The mean and standard deviation of diameters based on the collected data are $\mu_x = 1.200''$ and $\sigma_x = 0.005''$.

    (a) If the diameter of the crankshaft is specified as $1.200 \pm 0.020''$, calculate the percent of discarded crankshafts by using Microsoft Excel function.

    (b) Use the standard normal distribution table to calculate the probability $P(1.180 \leq X < 1.205)$,

    (c) Use the MATLAB command to calculate the probability:

$$P(1.200 - 3\sigma_x \leq X \leq 1.200 + 3\sigma_x)$$

$$P(1.200 - 2\sigma_x \leq X \leq 1.200 + 2\sigma_x)$$

$$P(1.200 - \sigma_x \leq X \leq 1.200 + \sigma_x).$$

2.46. Use the MATLAB to plot the PDFs and CDFs of a lognormal distribution with three different sets of distribution parameters: (1) $\mu_{\ln x} = 3.25$ and $\sigma_{\ln x} = 0.129$; (2) $\mu_{\ln x} = 3.25$ and $\sigma_{\ln x} = 0.516$; and (3) $\mu_{\ln x} = 3.25$ and $\sigma_{\ln x} = 1.125$. Discuss the effect of the standard deviation $\sigma_{\ln x}$ on the lognormal PDF.

2.47. Use Excel to plot the PDFs and CDFs of a lognormal distribution with three different sets of distribution parameters: (1) $\mu_{\ln x} = 1.25$ and $\sigma_{\ln x} = 0.129$; (2) $\mu_{\ln x} = 3.25$ and $\sigma_{\ln x} = 0.129$; and (3) $\mu_{\ln x} = 5.25$ and $\sigma_{\ln x} = 0.129$. Discuss the effect of the mean $\mu_{\ln x}$ on the lognormal PDF.

2.48. The fatigue strength X (ksi) of a material is a random variable and can be described by a lognormal distribution with the mean $\mu_{\ln x} = 3.25$ and the standard deviation $\sigma_{\ln x} = 0.189$.

   (a) Use MATLAB or Excel to plot its PDF and CDF.

   (b) Calculate the mean μ_x and the standard deviation σ_x of the fatigue strength.

   (c) Use MATLAB command to calculate probability $P\,(4 \leq X \leq 210)$ and $P\,(X \geq 10)$.

   (d) Use Excel to calculate probability $P\,(88 \leq X)$ and $P\,(X \geq 15)$.

2.49. Use MATLAB to plot the PDFs and the CDFs of a Weibull distribution with three different sets of distribution parameters: (1) $\eta = 1$ and $\beta = 0.5$; (2) $\eta = 1$ and $\beta = 1.0$; and (3) $\eta = 1$ and $\beta = 5$. Discuss the effect of shape parameter β on a Weibull PDF.

2.50. Use Excel to plot the PDFs and the CDFs of a Weibull distribution with three different sets of distribution parameters: (1) $\eta = 0.5$ and $\beta = 5$; (2) $\eta = 5$ and $\beta = 5$; and (3) $\eta = 10$ and $\beta = 5$. Discuss the effect of scale parameter η on the Weibull PDF.

2.51. The ultimate tensile strength X (ksi) of a type of steel is a random variable and follows a Weibull distribution with parameters $\beta = 20$, and $\eta = 100$.

   (a) Calculate the mean and standard deviation of the ultimate strength.

   (b) Use the Excel function to calculate the probability: $P\,(X \leq 105)$; $P\,(98 \leq X \leq 102)$.

   (c) Use the MATLAB command to calculate the probability: $P\,(20 \leq X \leq 80)$; $P\,(50 \leq X)$.

2.52. A two-speed synchronized transfer case used in a large industrial dump truck experiences failure that seems to be well approximated by a two-parameter Weibull distribution with $\eta = 18,000$ km and $\beta = 2.7$.

   (a) What is the probability $P(x > 10,000)$?

   (b) What is the probability $P(x > 24,000)$?

2.53. It is suggested that the stress x of a bridge connection was an exponential distribution with mean value 6 MPa.

   (a) The probability $P(x < 10)$.

   (b) The probability $P(5 < x < 10)$.

2.54. The time to failure T of a system is an exponential distribution with the parameter $\lambda = 0.001$. Calculate the probability of the system that can work more than 200 (h).

2.55. The time to failure X of a system is a random variable and can be described by an exponential distribution with the parameter $\lambda = 0.002$.

(a) Use MATLAB or Excel to plot its PDF and CDF.

(b) Calculate the probability of the system that can work more than 100 (h).

2.56. A set of 200 test data X of ultimate material strength is collected. The mean μ_x and the standard deviation σ_x calculated from the collected data are: $\mu_x = 101.53$ (ksi) and $\sigma_x = 55.37$ (ksi). The ten non-overlapping intervals and the observed number of data in the corresponding intervals are shown in Table 2.21.

Table 2.21: Intervals and observed number of data

Non-Overlapping Interval		o_j	e_j	$\dfrac{(o_j - e_j)^2}{e_j}$
Lower range x_{j-L}	Upper range x_{j-U}			
17.10	48.88	22		
48.88	80.66	64		
80.66	112.44	49		
112.44	144.22	30		
144.22	176.00	16		
176.00	207.78	8		
207.78	239.56	3		
239.56	271.34	5		
271.34	303.12	2		
303.12	334.91	1		
			$W = \displaystyle\sum_{j=1}^{10} \dfrac{(o_j - e_j)^2}{e_j}$	

Use Table 2.21 and the Excel function to do the following.

(a) Use the χ^2 test to check the assumption with a significant level $a = 0.05$ that it follows a normal distribution.

(b) Use the χ^2 test to check the assumption with a significant level $a = 0.05$ that it follows a lognormal distribution.

(c) Use the χ^2 test to check the assumption with a significant level $a = 0.05$ that it follows a Weibull distribution.

2.57. A yield strength data of material specimen are listed in Table 2.22. Input the data in the first column in an Excel file.

Table 2.22: Yield strength

Yield Strength Data
61.1, 68.5, 45.3, 63.0, 59.9, 50.7, 55.7, 60.0, 78.3, 73.7, 50.5, 75.2, 62.2, 57.7, 62.1, 56.9, 57.4, 66.5, 66.1, 66.1,
61.9, 51.3, 62.2, 67.3, 60.9, 63.9, 62.2, 56.4, 59.8, 53.7, 63.1, 51.6, 52.1, 53.5, 41.5, 66.2, 59.9, 53.8, 65.8, 48.4,
57.5, 56.7, 59.9, 59.9, 53.2, 57.9, 57.2, 61.6, 64.3, 64.4, 53.2, 58.5, 51.2, 51.8, 58.1, 66.8, 53.8, 60.2, 56.8, 64.4,
51.9, 58.3, 61.2, 64.3, 66.8, 58.6, 49.7, 53.9, 52.1, 71.4, 54.6, 62.3, 57.0, 63.1, 53.8, 50.2, 50.1, 60.9, 57.1, 57.0,
66.1, 59.7, 59.2, 67.1, 53.6, 62.0, 62.8, 56.7, 59.3, 51.5, 51.6, 58.7, 62.2, 72.7, 54.3, 59.2, 57.6, 47.2, 55.6, 48.0

Create MATLAB programs to do the following.

(a) Build its histogram.

(b) Calculate the mean and standard deviation of the collected data.

(c) Use the χ^2 test to check the assumption with a significant level $a = 0.05$ that it follows a normal distribution.

(d) Use the χ^2 test to check the assumption with a significant level $a = 0.05$ that it follows a lognormal distribution.

(e) Use the χ^2 test to check the assumption with a significant level $a = 0.05$ that it follows a Weibull distribution.

CHAPTER 3

Computational Methods for the Reliability of a Component

3.1 INTRODUCTION

In reliability-based mechanical design, the reliability replaces the traditional factor of safety and is used to establish the relationship between material mechanical properties, component geometric dimensions, and loading conditions. So, the methods for calculating reliability are one of the key issues for reliability-based mechanical design. This chapter will discuss how to calculate the reliability of a component under specified loading. The following six topics will be discussed.

- The limit state function for a mechanical component: this book focuses on mechanical component design. The limit state function describes a limit state of a component between safety and failure. It is a governing equation for calculating the reliability of a component.

- Reliability of component for two random variables. When limit state functions have only two random variables, the interference method can be used to calculate the reliability of a component.

- The First-Order Second-Moment (FOSM) method: when all random variables are normal distributions in a limit state function, the FOSM method can be used to calculate the reliability of a component for a linear limit state function or to estimate the reliability of a component for a nonlinear limit state function.

- The Hasofer-Lind (H-L) method: when a limit state function is a nonlinear function of all normal random variables, the H-L method can be used to calculate the reliability of a component.

- The Rachwitz-Fiessler (R-F) method: when a limit state function contains non-normal random variables, the R-F method can be used to calculate the reliability of a component.

- The Monte Carlo method: it is a numerical simulation technique and can be used to calculate the reliability of a component with a limit state function that contains any type of random variables.

3.2   LIMIT STATE FUNCTION

A mechanical component of a ductile material will yield when the maximum component stress is more than the material yield strength. Such status of the component is a failure. When the maximum component normal stress is less than the yield strength, the status of the component is safe. The critical status between the safe and the failure will be a limit state when the maximum component normal stressis equal to the material yield strength.

Limit state function.   For a general case, let S represent component strength index, which is a permissible or allowable parameter of a component in the safe status such as yield strength, ultimate strength, allowable deflection, and fatigue strength. Let Q represent component stress index, which is a component parameter induced by loading such as maximum normal stress, maximum shear stress, maximum Von-Mises stress, maximum deflection, and fatigue damage. The limit state function $g(S, Q)$ of a component is defined as:

$$g(S, Q) = S - Q = \begin{cases} > 0 & \text{Safe} \\ = 0 & \text{Limit state} \\ < 0 & \text{Failure.} \end{cases} \tag{3.1}$$

A limit state function is a mathematic function and will be the function of other design parameters such as material mechanical properties, component geometric dimensions, and loading. For a limit state function, it can describe three possible states of a component, as shown in Equation (3.1). When the value of a limit state function is more than zero, the component is safe. When the value of a limit state function is less than zero, the component is a failure. When the value of a limit state function is equal to zero, the component is in a limit state between the safe and the failure or can be called as on the surface of the limit state function: $g(S, Q) = 0$.

Per the definition of a limit state function, the reliability of a component will be:

$$R = P[g(S, Q) \geq 0] = P(S - Q \geq 0). \tag{3.2}$$

The probability of failure of a component will be:

$$F = 1 - R = P[g(S, Q) < 0] = P(S - Q < 0). \tag{3.3}$$

The limit state function of a component will be established by the failure theories of the problem under consideration. Corresponding failure theories for a component under static loading will be discussed in Chapter 4. Failure theories for a component under cyclic loading will be discussed in Chapter 1 in *Reliability-Based Mechanical Design*, Volume 2. Here, we will give several examples of limit state functions.

Example 3.1

A shaft with a diameter d (in) is subjected to a pure torque T (lb.in). The material is ductile material with a shear yield strength S_{sy}. If the maximum shear stress failure theory is used, establish the limit state function of this shaft.

Solution:

Per the maximum shear stress theory, the component strength index in this example will be the shear yield strength S_y, that is:

$$S = S_{sy}.$$

The component stress index in this example will be the shear stress induced by the pure torque and can be calculated by the following equation:

$$Q = \tau_{max} = \frac{T \times d/2}{\pi d^4/32} = \frac{16T}{\pi d^3}.$$

When the maximum shear stress failure theory is used, the limit state function of the shaft is:

$$g\left(S, Q\right) = S - Q = g\left(S_y, d, T\right) = S_{sy} - \frac{16T}{\pi d^3} = \begin{cases} > 0 & \text{Safe} \\ = 0 & \text{Limit state} \\ < 0 & \text{Failure.} \end{cases}$$

The limit state function $g\left(S_{sy}, d, T\right)$ is the function of three design parameters S_{sy}, d, and T. Typically, these parameters will be random variables. Therefore, the limit state function $g\left(S_{sy}, d, T\right)$ will also be a random variable. ∎

Example 3.2

A cantilever beam, as shown in Figure 3.1, is subjected to a concentrated load P at the free end. The beam is a constant rectangular bar with a length L and a height h and a width b. The material's Young's modulus is E. The design specification is that the allowable deflection at the free end B will be Δ. Establish the limit state function of this problem.

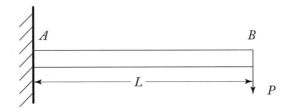

Figure 3.1: Schematic of a cantilever beam.

Solution:

In this example, the component strength index is the allowable deflection Δ, that is,

$$S = \Delta.$$

The component stress index will be the maximum deflection at the free end B. For a cantilever beam under a concentrated load at the free end, the maximum deflection at the free end is:

$$Q = \delta = \frac{PL^3}{3EI} = \frac{PL^3}{3E\left(\frac{1}{12}h^3b\right)} = \frac{4PL^3}{Eh^3b}.$$

According to the design requirement of this example, the limit state function will be:

$$g(S, Q) = S - Q = g(E, P, h, b, L) = \Delta - \frac{4PL^3}{Eh^3b} = \begin{cases} > 0 & \text{Safe} \\ = 0 & \text{Limit state} \\ < 0 & \text{Failure.} \end{cases}$$

The limit state function $g(E, P, h, b, L)$ is the function of five design parameters E, P, h, b, and L. ∎

Example 3.3

A beam structure is designed with an infinite fatigue life. The material fatigue endurance limit of the beam at the critical cross-section is S_e. The critical rectangular cross-section with a height h and a width b is subjected to a fully reversed cyclic bending moment with a magnitude M. Determine the limit state function of this beam structure at the critical cross-section.

Solution:

For this example, the component strength index will be the material fatigue endurance limit S_e. So, we have:

$$S = S_e.$$

The component stress index will be the stress amplitude of the fully reversed cyclic stress induced by a fully reversed cyclic bending moment. The following equation can calculate the component maximum bending stress:

$$Q = \sigma_a = \frac{M\left(\frac{h}{2}\right)}{I} = \frac{M\left(\frac{h}{2}\right)}{\frac{1}{12}h^3b} = \frac{6M}{h^2b}.$$

According to the fatigue theory, the limit state function of this example will be:

$$g(S, Q) = S - Q = g(S_e, h, b, M) = S_e - \frac{6M}{h^2b} = \begin{cases} > 0 & \text{Safe} \\ = 0 & \text{Limit state} \\ < 0 & \text{Failure.} \end{cases}$$

The limit state function $g\left(S_e, h, b, M\right)$ is the function of four parameters S_e, h, b, and M. ∎

3.3 RELIABILITY OF A COMPONENT WITH TWO RANDOM VARIABLES

When the limit state function of a component consists of only two mutually independent random variables, we can have an explicit formula for determining the reliability of a component. In this section, we will derive an explicit formula for a general case. Then, four special cases will be discussed, which are: both uniform distributions, both normal distributions, both log-normal distributions, and both exponential distributions.

3.3.1 INTERFERENCE METHOD

Let $f_S(s)$ and $F_S(s)$ represent the PDF and CDF of component strength index S. Let $f_Q(q)$ and $F_Q(q)$ represent the PDF and CDF of component stress index Q. In the following, we assume that S and Q are statistically independent.

The interference method is to use the following Equations (3.5) or (3.6) to calculate the reliability of a component when there are only two statistically independent random variables in a limit state function.

Now, we will derive these two equations. Based on Equation (3.2), the reliability of a component is

$$R = P\left[g\left(S, Q\right) \geq 0\right] = \iint\limits_{S-Q \geq 0} f_S(s) f_Q(q) ds dq = \iint\limits_{S \geq Q} f_S(s) f_Q(q) ds dq. \qquad (3.4)$$

If we run the integration concerning the random variable S first, as shown in Figure 3.2, the reliability of the component will become:

$$R = P\left[g\left(S, Q\right) \geq 0\right] = \iint\limits_{S \geq Q} f_S(s) f_Q(q) ds dq = \int_{-\infty}^{+\infty} f_Q(q) \left[\int_q^{+\infty} f_S(s) ds\right] dq.$$

$$= \int_{-\infty}^{+\infty} f_Q(q) \left[1 - \int_{-\infty}^q f_S(s) ds\right] dq = \int_{-\infty}^{+\infty} f_Q(q) \left[1 - F_S(q)\right] dq. \qquad (3.5)$$

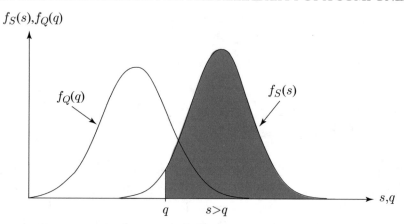

Figure 3.2: The schematic for integrating s first.

If we run the integration concerning the random variable Q first, as shown in Figure 3.3, the reliability of the component will become:

$$R = P\left[g\left(S, Q\right) \geq 0\right] = \iint_{S \geq Q} f_S\left(s\right) f_Q\left(q\right) ds\,dq = \int_{-\infty}^{+\infty} f_S\left(s\right) \left[\int_{-\infty}^{s} f_Q\left(q\right) dq\right] ds$$

$$= \int_{-\infty}^{+\infty} f_S\left(s\right) F_Q\left(s\right) ds. \tag{3.6}$$

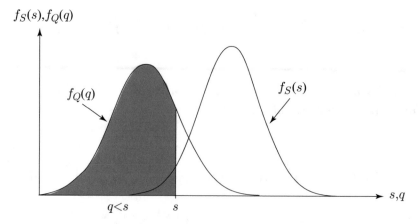

Figure 3.3: The schematic for integrating q first.

Equations (3.5) and (3.6) are the explicit formula for calculating the reliability of a component when the limit state function of a component contains only two random variables.

Example 3.4
For a bar under axial loading, the maximum normal stress follows a uniform distribution. Its PDF is:

$$f_\sigma(\sigma) = \begin{cases} \dfrac{1}{40} & 15 \text{ (ksi)} \leq \sigma \leq 55 \text{ (ksi)} \\ 0 & \text{otherwise.} \end{cases} \qquad (a)$$

The material yield strength also follows a uniform distribution with the following PDF:

$$f_{S_y}(s_y) = \begin{cases} \dfrac{1}{30} & 45 \text{ (ksi)} \leq s_y \leq 75 \text{ (ksi)} \\ 0 & \text{otherwise.} \end{cases} \qquad (b)$$

Determine the reliability of this bar under the specified axial loading.

Solution:
For this problem, the component strength index will be the material yield strength, and the component stress index will be the maximum normal stress induced by the axial loading. The limit state function will be:

$$g(S, Q) = S - Q = S_y - \sigma = \begin{cases} > 0 & \text{Safe} \\ = 0 & \text{Limit state} \\ < 0 & \text{Failure.} \end{cases} \qquad (c)$$

Then, we could use Equations (3.5) or (3.6) to calculate the reliability. In the following, we will use Equation (3.5) to calculate the reliability of the bar.

To use Equation (3.6) to run the calculation, we need to get the CDF of the yield strength. Based on the PDF specified in Equation (b), the CDF of the yield strength will be:

$$F_{S_y}(S_y) = \begin{cases} 0 & \sigma < 45 \text{ (ksi)} \\ \dfrac{S_y - 45}{30} & 45 \text{ (ksi)} \leq \sigma \leq 75 \text{ (}ksi\text{)} \\ 1 & \sigma > 75 \text{ (ksi).} \end{cases} \qquad (d)$$

Based on Equation (3.5), we have:

$$
\begin{aligned}
R &= \int_{-\infty}^{+\infty} f_\sigma\left(\sigma\right)\left[1 - F_{S_y}\left(\sigma\right)\right] d\sigma \\
&= \int_{-\infty}^{15} f_\sigma\left(\sigma\right)\left[1 - F_{S_y}\left(\sigma\right)\right] d\sigma + \int_{15}^{45} f_\sigma\left(\sigma\right)\left[1 - F_{S_y}\left(\sigma\right)\right] d\sigma \\
&\quad + \int_{45}^{55} f_\sigma\left(\sigma\right)\left[1 - F_{S_y}\left(\sigma\right)\right] d\sigma + \int_{55}^{75} f_\sigma\left(\sigma\right)\left[1 - F_{S_y}\left(\sigma\right)\right] d\sigma \\
&\quad + \int_{75}^{+\infty} f_\sigma\left(\sigma\right)\left[1 - F_{S_y}\left(\sigma\right)\right] d\sigma \\
&= \int_{-\infty}^{15} 0 \times [1-0]\, d\sigma + \int_{15}^{45} \frac{1}{40}[1-0]\, d\sigma + \int_{45}^{55} \frac{1}{40}\left[1 - \frac{\sigma-45}{30}\right] d\sigma \\
&\quad + \int_{55}^{75} 0 \times \left[1 - \frac{\sigma-45}{30}\right] d\sigma + \int_{75}^{+\infty} 0 \times [1-0)]\, d\sigma \\
&= \int_{15}^{45} \frac{1}{40}[1-0]\, d\sigma + \int_{45}^{55} \frac{1}{40}\left[1 - \frac{\sigma-45}{30}\right] d\sigma = \frac{1}{40}\sigma\Big|_{15}^{45} + \frac{1}{40}\left(\frac{75}{30}\sigma - \frac{1}{60}\sigma^2\right)\Big|_{45}^{55} \\
&= 0.75 + 0.2083 = 0.9583.
\end{aligned}
$$

■

3.3.2   COMPUTATION OF RELIABILITY WHEN BOTH ARE NORMAL DISTRIBUTIONS

When both component strength index S and the component stress index Q are normal distribution random variables, we can directly calculate the reliability of the component by using the standard normal distribution table, an Excel function, or a MATLAB command.

Let μ_S and σ_S represent the mean and standard deviation of the normally distributed component strength index S. Let μ_Q and σ_Q represent the mean and standard deviation of the normally distributed component stress index Q. Since the component strength index S and the component stress index Q are statistically independent and normally distributed random variables, the limit state function $g(S, Q) = S - Q$ will be a normally distributed random variable too. Based on Equation (2.83)–(2.87), the mean μ_g and the standard deviation σ_g of the normally distributed limit state function $g(S, Q)$ are determined by the following equations:

$$
\mu_g = \mu_S - \mu_Q \tag{3.7}
$$

$$
\sigma_g = \sqrt{\sigma_S^2 + \sigma_Q^2}. \tag{3.8}
$$

Since the limit state function $g(S, Q)$ is a normally distributed random variable, the reliability of the component can be calculated by the following equation if the Excel function is used:

$$R = P[g(S, Q) \geq 0] = 1 - P[g(S, Q) < 0]$$
$$= 1 - NORM.DIST(0, \mu_g, \sigma_g, true). \quad (3.9)$$

The reliability of the component can be calculated by the following equation if the MATLAB command is used:

$$R = P[g(S, Q) \geq 0] = 1 - P[g(S, Q) < 0] = 1 - normcdf(0, \mu_g, \sigma_g,). \quad (3.10)$$

If the standard normal distribution table is used, the reliability of the component can be calculated by the following equation:

$$R = P[g(S, Q) \geq 0] = 1 - P[g(S, Q) < 0] = 1 - \Phi\left(\frac{-\mu_g}{\sigma_g}\right).$$

$$= \Phi\left(\frac{\mu_g}{\sigma_g}\right) = \Phi\left(\frac{\mu_S - \mu_Q}{\sqrt{\sigma_S^2 + \sigma_Q^2}}\right) = \Phi(\beta). \quad (3.11)$$

The term $\dfrac{\mu_g}{\sigma_g}$ or $\dfrac{\mu_S - \mu_Q}{\sqrt{\sigma_S^2 + \sigma_Q^2}}$ will be frequently shown in a lot of formula or iterative process. We can use β to represent it, that is:

$$\beta = \frac{\mu_g}{\sigma_g} = \frac{\mu_S - \mu_Q}{\sqrt{\sigma_S^2 + \sigma_Q^2}}. \quad (3.12)$$

β is called the reliability index, which can be directly used to calculate the reliability, as shown in Equation (3.11).

Example 3.5
The failure of a component will be a static failure of brittle material. According to the design information, the component material ultimate strength S_u follows a normal distribution with a mean $\mu_{S_u} = 62.5$ (ksi) and a standard deviation $\sigma_{S_u} = 5.8$ (ksi). The maximum normal stress σ_{max} at the critical section follows a normal distribution with a mean $\mu_{\sigma_{max}} = 30.3$ (ksi) and a standard deviation $\sigma_{\sigma_{max}} = 15.4$ (ksi). Calculate the reliability of this component.

Solution:
In this example with a component of a brittle material, the component strength index is the ultimate material strength S_u and the component stress index is the maximum normal stress

$\sigma_{\max}$ in the critical section. Therefore, the limit state function of this component will be:

$$g\,(S_u, \sigma_{\max}) = S_u - \sigma_{\max} = \begin{cases} > 0 & \text{Safe} \\ = 0 & \text{Limit state} \\ < 0 & \text{Failure.} \end{cases} \tag{a}$$

Since both S_u and $\sigma_{\max}$ are normally distributed random variables and the limit state function $g\,(S_u, \sigma_{\max})$ is a linear function of S_u and $\sigma_{\max}$, the limit state function will be a normally distributed random variable too. The mean μ_g and the standard deviation σ_g can be calculated by Equations (3.7) and (3.8):

$$\mu_g = \mu_{S_u} - \mu_{\sigma_{\max}} = 62.5 - 30.3 = 32.2 \text{ (ksi)} \tag{b}$$

$$\sigma_g = \sqrt{\sigma_{S_u}^2 + \sigma_{\sigma_{\max}}^2} = \sqrt{5.8^2 + 15.4^2} = 16.456. \tag{c}$$

The reliability of the component can be calculated by Equations (3.9), (3.10), or (3.11). Use Equation (3.11) to run the calculation. In this example, the reliability index β per Equation (3.12) is:

$$\beta = \frac{\mu_g}{\sigma_g} = \frac{32.2}{16.456} = 1.9567. \tag{d}$$

Per Equation (3.11) and Table 2.10, the reliability of the component is:

$$R = \Phi\,(\beta) = \Phi\,(1.9567) = 0.9748.$$

■

3.3.3   COMPUTATION OF RELIABILITY WHEN BOTH ARE LOG-NORMAL DISTRIBUTIONS

When both component strength index S and component stress index Q follow lognormal distributions, the event $(S > Q)$ is the same as the event $(\ln S > \ln Q)$ because both S and Q are always positive. Therefore, for this case, the limit state function of a component can be expressed as

$$g\,(S, Q) = \ln S - \ln Q = \begin{cases} > 0 & \text{Safe} \\ = 0 & \text{Limit state} \\ < 0 & \text{Failure.} \end{cases} \tag{3.13}$$

Now, since $\ln S$ and $\ln Q$ are normal distributions, the limit state function $g\,(S, Q)$ will be a normal distribution too. Per Equations (3.7) and (3.8), the mean and standard deviation of the limit state function in this case are:

$$\mu_g = \mu_{\ln S} - \mu_{\ln Q} \tag{3.14}$$

$$\sigma_g = \sqrt{\sigma_{\ln S}^2 + \sigma_{\ln Q}^2}, \tag{3.15}$$

where $\mu_{\ln S}$ and $\sigma_{\ln S}$ are the log-mean and log-standard deviation of the log-normally distributed component strength index S. $\mu_{\ln Q}$ and $\sigma_{\ln Q}$ are the log-mean and log-standard deviation of the log-normally distributed component stress index Q.

Equations (3.9), (3.10), or (3.11) can be used to calculate the reliability of a component when both are log-normal distributions.

Example 3.6
The failure mode of a component will be a fatigue failure. According to the design information, at the specified cyclic stress level, both the number of cycles to failure N_f of the component and the service number of cycles n_s are log-normal distributions. The log mean and log standard deviation of N_f are $\mu_{\ln N_f} = 13.305$ and $\sigma_{\ln N_f} = 0.121$. The log mean and log standard deviation of n_s are $\mu_{\ln n_s} = 11.886$ and $\sigma_{\ln n_s} = 0.654$. Calculate the reliability of this component.

Solution:
Per the provided information, the component strength index S will be the number of cycles to failure N_f of the component at the specified cyclic stress level. The component stress index Q will be the service number of cycles n_s at the specified cyclic stress level. So, the limit state function of the component is:

$$g\left(N_f, n_s\right) = N_f - n_s = \begin{cases} > 0 & \text{Safe} \\ = 0 & \text{Limit state} \\ < 0 & \text{Failure.} \end{cases} \tag{a}$$

Since both N_f and n_s are positive variables, the limit state function of this component can be expressed equivalently by the following equation. We will use the following equation to solve this example:

$$g\left(N_f, n_s\right) = \ln(N_f) - \ln(n_s) = \begin{cases} > 0 & \text{Safe} \\ = 0 & \text{Limit state} \\ < 0 & \text{Failure.} \end{cases} \tag{b}$$

Per Equations (3.14) and (3.15), we have the mean μ_g and standard deviation σ_g of the limit state function:

$$\mu_g = \mu_{\ln(N_f)} - \mu_{\ln(n_s)} = 13.305 - 11.886 = 1.419 \tag{c}$$

$$\sigma_g = \sqrt{\sigma_{\ln(N_f)}^2 + \sigma_{\ln(n_s)}^2} = \sqrt{0.121^2 + 0.654^2} = 0.6651. \tag{d}$$

We can use any one of Equations (3.9), (3.10), and (3.11) to calculate the reliability of the component. Use Equation (3.9).

$$R = P\left[g\left(N_f, n_s\right) \geq 0\right]$$
$$= 1 - NORM.DIST\left(0, 1.419, 0.6651, true\right) = 1 - 0.01644 = 0.9836.$$

∎

3.3.4   COMPUTATION OF RELIABILITY WHEN BOTH ARE EXPONENTIAL DISTRIBUTIONS

When both component strength index S and component stress index Q follow exponential distributions, we can also use the interference method to calculate the reliability of the component. The exponentially distributed component strength index S with the distribution parameter λ_S has the PDF and the CDF as

$$f_S(s) = \lambda_S e^{-\lambda_S S} \qquad 0 \leq s \leq \infty \tag{3.16}$$
$$F_S(s) = 1 - e^{-\lambda_S S} \qquad 0 \leq s \leq \infty. \tag{3.17}$$

The exponentially distributed component strength index Q with the distribution parameter λ_Q has the PDF as

$$f_Q(q) = \lambda_Q e^{-\lambda_Q q} \qquad 0 \leq q \leq \infty \tag{3.18}$$
$$F_Q(q) = 1 - e^{-\lambda_Q q} \qquad 0 \leq q \leq \infty. \tag{3.19}$$

Equations (3.5) or (3.6) can be used to calculate the reliability of the component. Use Equation (3.6) to run the calculation:

$$R = P\left[g\left(S, Q\right) \geq 0\right] = \int_0^{+\infty} f_S(s) F_Q(s)\, ds = \int_0^{+\infty} \lambda_S e^{-\lambda_S S}\left[1 - e^{-\lambda_Q s}\right] ds$$
$$= \int_0^{+\infty}\left[\lambda_S e^{-\lambda_S S} - \lambda_S e^{-(\lambda_S + \lambda_Q)s}\right] ds = \left[-e^{-\lambda_S S} + \frac{\lambda_S}{\lambda_S + \lambda_Q} e^{-(\lambda_S + \lambda_Q)s}\right]\Bigg|_0^{\infty}$$
$$= 1 - \frac{\lambda_S}{\lambda_S + \lambda_Q} = \frac{\lambda_Q}{\lambda_S + \lambda_Q}. \tag{3.20}$$

Example 3.7

The failure mode of a component will be a static failure when the maximum stress at the critical section is more than the ultimate material strength. It is assumed that both the maximum stress σ_{max} and the ultimate strength S_u follow an exponential distribution with a mean $\mu_{\sigma_{max}} = 12.4$ (ksi) and a mean $\mu_{S_u} = 62.5$ (ksi). Calculate the reliability of this component.

Solution:

Based on the provided information in this problem, the component strength index is the ultimate material strength S_u and the component stress index is the maximum stress $\sigma_{\max}$ at the critical section. The limit state function of this component will be:

$$g\left(S_u, \sigma_{\max}\right) = S_u - \sigma_{\max} = \begin{cases} > 0 & \text{Safe} \\ = 0 & \text{Limit state} \\ < 0 & \text{Failure.} \end{cases}$$

Per the provided information, the distribution parameters of the exponentially distributed ultimate strength, and the exponentially distributed maximum stress are:

$$\lambda_{S_u} = \frac{1}{\mu_{S_u}} = \frac{1}{62.5} = 0.016$$

$$\lambda_{\sigma_{\max}} = \frac{1}{\mu_{\sigma_{\max}}} = \frac{1}{12.4} = 0.081.$$

Per Equation (3.20), the reliability of this component is:

$$R = P\left[g\left(S_u, \sigma_{\max}\right) \geq 0\right] = \frac{\lambda_{\sigma_{\max}}}{\lambda_{S_u} + \lambda_{\sigma_{\max}}} = \frac{0.081}{0.016 + 0.081} = 0.835.$$

■

3.4   RELIABILITY INDEX β

The reliability index β has been mentioned in Section 3.2 and defined by Equation (3.12):

$$\beta = \frac{\mu_g}{\sigma_g} = \frac{\mu_S - \mu_Q}{\sqrt{\sigma_S^2 + \sigma_Q^2}}. \tag{3.12}$$

The reliability index β will be a frequently used important parameter, which can be used to calculate the reliability of the component per Equation (3.11):

$$R = P\left[g\left(S, Q\right) \geq 0\right] = \Phi\left(\frac{\mu_g}{\sigma_g}\right) = \Phi(\beta). \tag{3.11}$$

Now, we will explain the physical meaning of the reliability index β.

In the limit state function of a component, the component strength index S and the component stress index Q are random variables and can be functions of other random variables. For a general case, let use the following general expression of the surface of a limit state function:

$$g\left(S, Q\right) = S - Q = g\left(X_1, X_2, \ldots, X_n\right) = 0. \tag{3.21}$$

This limit state function can be a linear function and a nonlinear function of all random variables. If all random variables $X_i (i = 1, 2, \ldots, n)$ in the limit state function are statistically independent normal distributions, we can convert them into statistically independent standard normally distributed random variables $Z_i (i = 1, 2, \ldots, n)$ by the following equation:

$$Z_i = \frac{X_i - \mu_{X_i}}{\sigma_{X_i}} \qquad i = 1, 2, \ldots, n, \tag{3.22}$$

where μ_{X_i} and σ_{X_i} are the mean and standard deviation of a normally distributed random variable X_i. Z_i is a corresponding standard normal distributed random variable of the normal distributed random variable X_i. After these conversions, the surface of the limit state function (3.21) becomes:

$$g(S, Q) = S - Q = g(X_1, X_2, \ldots, X_n) = g_1(Z_1, Z_2, \ldots, Z_n) = 0. \tag{3.23}$$

$g_1(Z_1, Z_2, \ldots, Z_n) = 0$ is a surface specified in the standard normally distributed space (a coordinate system), which consists of standard normally distributed random variables $Z_i (i = 1, 2, \ldots, n)$.

The reliability index β is the shortest distance [1, 2] from the origin of the standard normally distributed space to the surface of the limit state function: $g_1(Z_1, Z_2, \ldots, Z_n) = 0$.

Example 3.8

The limit state function of a component is $g(S, Q) = S - Q$. Both S and Q are normal distributions. The component strength index S has a mean μ_S and a standard deviation σ_S. The component stress index Q has a mean μ_Q and a standard deviation σ_Q. Use this limit state function to verify that the shortest distance between the origin of the standard normally distributed space and the surface of the limit state function $g(S, Q) = S - Q = 0$ is the reliability index $\beta = (\mu_S - \mu_Q)/\sqrt{\sigma_S^2 + \sigma_Q^2}$.

Solution:

The surface of the limit state function $g(S, Q) = S - Q$ is:

$$g(S, Q) = S - Q = 0. \tag{a}$$

In the S vs. Q coordinate system where the horizontal axis is Q, and the vertical axis is S, the limit state function will be a straight line, as shown in Figure 3.4. The region on the left side of the line of the limit state function is safe because of $S > Q$. The region on the right side of the line of the limit state function is a failure region because of $S < Q$.

Let us convert the component strength index S and the component stress index Q into standard normally distributed random variables Z_S and Z_Q:

$$Z_S = \frac{S - \mu_S}{\sigma_S}, \qquad Z_Q = \frac{Q - \mu_Q}{\sigma_Q}. \tag{b}$$

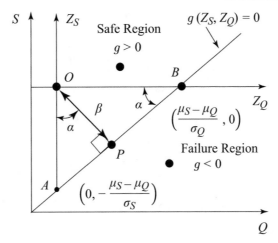

Figure 3.4: The line of a limit state function with two normal distributed random variables.

Rewrite Equation (b) as:

$$S = Z_S \times \sigma_S + \mu_S, \qquad Q = Z_Q \times \sigma_Q + \mu_Q. \tag{c}$$

Plug Equation (c) into Equation (a), we have a different expression of the surface of the same limit state function:

$$g_1 (Z_S, Z_Q) = (Z_S \times \sigma_S + \mu_S) - (Z_Q \times \sigma_Q + \mu_Q) = 0. \tag{d}$$

The standard normal distribution space, in this case, is the Z_S vs. Z_Q coordinate system. At point A with $Z_{Q-A} = 0$, Z_S can be obtained from Equation (d):

$$(Z_{S-A} \times \sigma_S + \mu_S) - (0 \times \sigma_Q + \mu_Q) = 0. \tag{e}$$

Therefore,

$$Z_{S-A} = -\frac{\mu_S - \mu_Q}{\sigma_S}. \tag{f}$$

So, the coordinate values at the point A in the standard normal distribution space is $\left(0, -\frac{\mu_S - \mu_Q}{\sigma_S}\right)$, as shown in Figure 3.4. Repeating the same procedure, we can get the coordinate values at the point B as $\left(\frac{\mu_S - \mu_Q}{\sigma_Q}, 0\right)$, as shown in Figure 3.4.

The line OP, as shown in Figure 3.4, is normal to the surface (line) of the limit state function. So, the length of OP will be the shortest distance between the origin O and the line of the limit state function in the standard normal distribution space. Therefore, the length of OP should be the reliability index β.

From Figure 3.4, in the triangle OPB, we have:

$$\sin(a) = \frac{\beta}{\frac{\mu_S - \mu_Q}{\sigma_Q}} = \frac{\beta \times \sigma_Q}{\mu_S - \mu_Q}. \tag{g}$$

From Figure 3.4, in the triangle OPA, we have:

$$\cos(a) = \frac{\beta}{\frac{\mu_S - \mu_Q}{\sigma_S}} = \frac{\beta \times \sigma_S}{\mu_S - \mu_Q}. \tag{h}$$

Square both sides of the Equations (g) and (h) and then add them together, we have

$$\sin^2(a) + \cos^2(a) = \left(\frac{\beta \times \sigma_Q}{\mu_S - \mu_Q} \right)^2 + \left(\frac{\beta \times \sigma_S}{\mu_S - \mu_Q} \right)^2. \tag{i}$$

Rewrite Equation (i), we can obtain:

$$\beta = \frac{\mu_S - \mu_Q}{\sqrt{\sigma_S^2 + \sigma_Q^2}}. \tag{j}$$

This result verifies that the shortest distance between the origin of the standard normally distributed space and the surface of the limit state function is equal to $(\mu_S - \mu_Q)/\sqrt{\sigma_S^2 + \sigma_Q^2}$ which is the value of the reliability index β. ∎

3.5   THE FIRST-ORDER SECOND-MOMENT (FOSM) METHOD

For a limit state function containing more than two random variables, if all random variables are normally distributed random variables, we can use the First-Order Second-Moment (FOSM) method to calculate its reliability.

The First-Order Second-Moment (FOSM) method is a probabilistic method to determine the reliability of a component when its limit state function is simplified by a first-order Taylor series. Then, the first and second moments (mean and standard deviation) of all random variables are used for the calculation of the reliability.

When the type of distribution of a random variable is unknown due to the lack of data, we can use its mean and standard deviation (the first and second moments) for estimating its reliability by using the FOSM method. This result will be only a rough approximate result of the reliability. In this section, we will only discuss the following two cases.

- The FOSM method for a linear limit state function: when a limit state function is a linear function of all normally distributed random variables, the FOSM method will provide an accurate result of the reliability of a component.

- The FOSM method for a nonlinear limit state function. When a limit state function is a nonlinear function of all normally distributed random variables, the FOSM method will only provide an approximate result of the reliability of a component.

3.5.1   THE FOSM METHOD FOR A LINEAR LIMIT STATE FUNCTION

A limit state function of a component, which is a linear function of the statistically independent normally distributed random variables, can be express as:

$$g(X_1, X_2, \ldots, X_n) = a_0 + \sum_{i=1}^{n} (a_i X_i) = \begin{cases} > 0 & \text{Safe} \\ = 0 & \text{Limit state} \\ < 0 & \text{Failure,} \end{cases} \tag{3.24}$$

where $a_i \, (i = 0, 1, \ldots, n)$ is a constant. $X_i \, (i = 1, 2, \ldots, n)$ is a normally distributed random variable with a mean μ_{X_i} and a standard deviation σ_{X_i}.

Since all random variables are statistically independent normally distributed random variables, the limit state function $g(X_1, X_2, \ldots, X_n)$ will be a normally distributed random variable. The mean μ_g and the standard deviation σ_g of the limit state function will be:

$$\mu_g = a_0 + \sum_{i=1}^{n} (a_i \mu_{X_i}) \tag{3.25}$$

$$\sigma_g = \sqrt{\sum_{i=1}^{n} (a_i \sigma_{X_i})^2}. \tag{3.26}$$

The reliability index β and the reliability R of the component will be:

$$\beta = \frac{\mu_g}{\sigma_g} = \frac{a_0 + \sum_{i=1}^{n} (a_i \mu_{X_i})}{\sqrt{\sum_{i=1}^{n} (a_i \sigma_{X_i})^2}} \tag{3.27}$$

$$R = \Phi(\beta) = \Phi\left(\frac{\mu_g}{\sigma_g}\right) = \Phi\left(\frac{a_0 + \sum_{i=1}^{n} (a_i \mu_{X_i})}{\sqrt{\sum_{i=1}^{n} (a_i \sigma_{X_i})^2}}\right). \tag{3.28}$$

Example 3.9
The component fails when its maximum normal stress is more than the ultimate material strength. The ultimate material strength S_u of this component is a normally distributed random variable with a mean $\mu_{S_u} = 61.5$ (ksi) and a standard deviation $\sigma_{S_u} = 5.91$ (ksi). The maximum normal stress at the critical section is the sum of normal stress σ_B by bending moment and the normal stress σ_A by axial loading. σ_B is a normally distributed random variable with a mean

$\mu_{\sigma_B} = 36.3$ (ksi) and a standard deviation $\sigma_{\sigma_B} = 4.21$ (ksi). σ_A is a normally distributed random variable with a mean $\mu_{\sigma_A} = 6.81$ (ksi) and a standard deviation $\sigma_{\sigma_A} = 2.65$ (ksi). Calculate the reliability of this component.

Solution:

In this example, the component strength index will be the ultimate material strength S_u. The component stress index will be the sum of normal stress σ_B by bending moment and the normal stress σ_A by axial loading. Therefore, the limit state function of this component will be:

$$g(S_u, \sigma_B, \sigma_A) = S_u - (\sigma_B + \sigma_A) = \begin{cases} > 0 & \text{Safe} \\ = 0 & \text{Limit state} \\ < 0 & \text{Failure.} \end{cases} \tag{a}$$

Since all three random variables S_u, σ_B, and σ_A are normally distributed random variables, and the limit state function is a linear function of them, we can use Equations (3.25) and (3.26) to calculate the mean and standard deviation of the limit state function. They are:

$$\mu_g = \mu_u - \mu_{\sigma_B} - \mu_{\sigma_A} = 61.5 - 36.3 - 6.81 = 18.39 \ \text{(ksi)} \tag{b}$$

$$\sigma_g = \sqrt{\sigma_u^2 + \sigma_{\sigma_B}^2 + \sigma_{\sigma_A}^2} = \sqrt{5.91^2 + 4.21^2 + 2.65^2} = 7.7248. \tag{c}$$

Per Equation (3.27), we can calculate the reliability index β:

$$\beta = \frac{\mu_g}{\sigma_g} = \frac{18.39}{7.7248} = 2.3806. \tag{d}$$

Per Equation (3.28), we can calculate the reliability of the component. We can use Excel to calculate the reliability of this component

$$R = \Phi(2.3806) = NORM.DIST(2.3806, 0, 1, true) = 0.9914.$$

■

3.5.2   THE FOSM METHOD FOR A NONLINEAR STATE FUNCTION

When a limit state function is a nonlinear function of all normally distributed random variable, we can obtain an approximate value of the reliability by using the FOSM method. In the FOSM method, the nonlinear limit state function will be simplified by the first-order Taylor series expansion at the mean point where all random variables have their means. The first-order Taylor series expansion at the mean point will be:

$$g\left(X_1, X_2, \ldots, X_n\right) = g\left(\mu_{X_1}, \mu_{X_2}, \ldots, \mu_{X_n}\right) + \sum_{i=1}^{n} G_{xi} \times \left(X_i - \mu_{X_i}\right)$$

$$= \begin{cases} > 0 & \text{Safe} \\ = 0 & \text{Limit state} \\ < 0 & \text{Failure,} \end{cases} \tag{3.29}$$

where, μ_{X_i} is the mean of the normally distributed random variable X_i, $(i = 1, 2, \ldots, n)$. The mean point is the point where every variable takes its mean value, that is, $\left(\mu_{X_1}, \mu_{X_2}, \ldots, \mu_{X_n}\right)$.

G_{xi} is a Taylor series constant-coefficient for Taylor expansion term $\left(X_i - \mu_{X_i}\right)$. It is calculated by the following equation at the mean point:

$$G_{xi} = \left. \frac{\partial g\left(X_1, X_2, \ldots, X_n\right)}{\partial X_i} \right|_{\text{at the mean point}} \qquad i = 1, 2, \ldots, n. \tag{3.30}$$

Now the simplified limit state function in Equation (3.29) is a linear function of all normal distributed random variables. We can use Equations (3.25) and (3.26) to calculate the mean and standard deviation of the simplified limit state function:

$$\mu_g = g\left(\mu_{X_1}, \mu_{X_2}, \ldots, \mu_{X_n}\right) \tag{3.31}$$

$$\sigma_g = \sqrt{\sum_{i=1}^{n} \left(G_{xi} \times \sigma_{X_i}\right)^2}. \tag{3.32}$$

Per Equations (3.27) and (3.28), the reliability index β, and the reliability R of the component will be:

$$\beta = \frac{\mu_g}{\sigma_g} = \frac{g\left(\mu_{X_1}, \mu_{X_2}, \ldots, \mu_{X_n}\right)}{\sqrt{\sum_{i=1}^{n} \left(G_{xi} \times \sigma_{X_i}\right)^2}} \tag{3.33}$$

$$R = \Phi(\beta) = \Phi\left(\frac{g\left(\mu_{X_1}, \mu_{X_2}, \ldots, \mu_{X_n}\right)}{\sqrt{\sum_{i=1}^{n} \left(G_{xi} \times \sigma_{X_i}\right)^2}} \right). \tag{3.34}$$

Here are some comments about the FOSM method for a nonlinear limit state function.

- The FOSM method will provide an approximate value of the reliability of a component when the limit state function is a nonlinear function of all normally distributed random variables.

- Since there might have different versions of a limit state function for the same problem, we might get a different estimation of the reliability of the same problem [2]. We will use the following example to demonstrate and to explain this situation.

Example 3.10

A simple support beam is under a concentrated loading at the middle of a beam, as shown in Figure 3.5. The yield strength S_y, the concentrated load P, the beam span L and the section modulus Z of the beam are all normally distributed random variables. Their distribution parameters are: $\mu_{S_y} = 6 \times 10^5$ (kN/m^2), $\sigma_{S_y} = 10^5$ (kN/m^2); $\mu_P = 10$ (kN), $\sigma_P = 2$ (kN); $\mu_L = 8$ (m), $\sigma_L = 2.083 \times 10^{-2}$ (m); and $\mu_Z = 10^{-4}$ (m^3), $\sigma_{S_y} = 2 \times 10^{-5}$ (m^3). Use the FOSM method to calculate the reliability of the beam by using the following two different limit state functions.

1. The yield strength is the component strength index. So, the limit state function is:

$$g\left(S_y, P, L, Z\right) = S_y - PL/(4Z) = \begin{cases} > 0 & \text{Safe} \\ = 0 & \text{Limit state} \\ < 0 & \text{Failure.} \end{cases} \tag{a}$$

2. The allowable bending moment is the component strength index. So, the limit state function is:

$$g\left(S_y, P, L, Z\right) = S_y Z - PL/4 = \begin{cases} > 0 & \text{Safe} \\ = 0 & \text{Limit state} \\ < 0 & \text{Failure.} \end{cases} \tag{b}$$

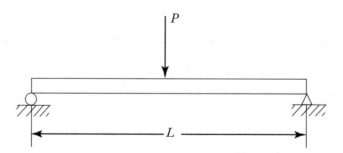

Figure 3.5: A simple support beam with a concentrated loading.

Solution:

In this example, all related random variables are normal distributions. However, both limit state functions are a nonlinear function of these random variables. We can use Equations (3.30)–(3.34) to conduct the calculations.

(1) The reliability of the beam based on the limit state function: $g\left(S_y, P, L, Z\right) = S_y - \dfrac{PL}{4Z}$.

Per Equation (3.31), we can calculate the mean of the limit state function:

$$\mu_g = g\left(\mu_{S_y}, \mu_P, \mu_L, \mu_Z\right) = \mu_{S_y} - \frac{\mu_P \mu_L}{4\mu_Z} = 6 \times 10^5 - \frac{10 \times 8}{4 \times 10^{-4}} = 4 \times 10^5. \qquad (c)$$

Per Equation (3.30), we can calculate the Tylor series constant coefficients:

$$G_{S_y} = \left.\frac{\partial g\left(S_y, P, L, Z\right)}{\partial S_y}\right|_{\text{at the mean point}} = 1$$

$$G_P = \left.\frac{\partial g\left(S_y, P, L, Z\right)}{\partial P}\right|_{\text{at the mean point}} = \left.-\frac{L}{4Z}\right|_{\text{at the mean point}} = -\frac{8}{4 \times 10^{-4}} = -2 \times 10^4$$

$$G_L = \left.\frac{\partial g\left(S_y, P, L, Z\right)}{\partial L}\right|_{\text{at the mean point}} = \left.-\frac{P}{4Z}\right|_{\text{at the mean point}} = -\frac{10}{4 \times 10^{-4}} = -2.5 \times 10^4$$

$$G_Z = \left.\frac{\partial g\left(S_y, P, L, Z\right)}{\partial Z}\right|_{\text{at the mean point}} = \left.\frac{PL}{4Z^2}\right|_{\text{at the mean point}} = \frac{10 \times 8}{4 \times \left(10^{-4}\right)^2} = 2 \times 10^9.$$

Per Equation (3.32), we can calculate the standard deviation of the limit state function:

$$\sigma_g = \sqrt{\left(G_{S_y} \times \sigma_{S_y}\right)^2 + \left(G_P \times \sigma_P\right)^2 + \left(G_L \times \sigma_L\right)^2 + \left(G_Z \times \sigma_Z\right)^2}$$

$$= \sqrt{\begin{array}{c}\left(1 \times 10^5\right)^2 + \left(-2 \times 10^4 \times 2\right)^2 + \left(-2.5 \times 10^4 \times 2.083 \times 10^{-2}\right)^2 \\ + \left(2 \times 10^9 \times 2 \times 10^{-5}\right)^2\end{array}}$$

$$= 114892.4331. \qquad (d)$$

Per Equation (3.33) and data in Equations (c) and (d), the reliability index β is:

$$\beta = \frac{\mu_g}{\sigma_g} = \frac{4 \times 10^5}{114892.4331} = 3.4815. \qquad (e)$$

Per Equation (3.34) and data in Equation (e), we have the reliability R with the use of MATLAB command:

$$R = \Phi\left(\beta\right) = \Phi\left(3.4815\right) = normcdf\left(3.4815, 0, 1\right) = 0.9998.$$

(2) The reliability of the beam based on the limit state function: $g\left(S_y, P, L, Z\right) = S_y Z - \dfrac{PL}{4}$.

We can repeat the above steps.

Per Equation (3.31), the mean of the limit state function is

$$\mu_g = g\left(\mu_{S_y}, \mu_P, \mu_L, \mu_Z\right) = \mu_{S_y}\mu_Z - \frac{\mu_P\mu_L}{4} = 6 \times 10^5 \times 10^{-4} - \frac{10 \times 8}{4} = 40. \quad \text{(f)}$$

Per Equation (3.30), we can calculate the Taylor series constant coefficients:

$$G_{S_y} = \left.\frac{\partial g\left(S_y, P, L, Z\right)}{\partial S_y}\right|_{\text{at the mean point}} = Z|_{\text{at the mean point}} = 10^{-4}$$

$$G_P = \left.\frac{\partial g\left(S_y, P, L, Z\right)}{\partial P}\right|_{\text{at the mean point}} = \left.-\frac{L}{4}\right|_{\text{at the mean point}} = -\frac{8}{4} = -2$$

$$G_L = \left.\frac{\partial g\left(S_y, P, L, Z\right)}{\partial L}\right|_{\text{at the mean point}} = \left.-\frac{P}{4}\right|_{\text{at the mean point}} = -\frac{10}{4} = -2.5$$

$$G_Z = \left.\frac{\partial g\left(S_y, P, L, Z\right)}{\partial Z}\right|_{\text{at the mean point}} = S_y|_{\text{at the mean point}} = 6 \times 10^5.$$

Per Equation (3.32), we can calculate the standard deviation of the limit state function

$$\sigma_g = \sqrt{\left(G_{S_y} \times \sigma_{S_y}\right)^2 + \left(G_P \times \sigma_P\right)^2 + \left(G_L \times \sigma_L\right)^2 + \left(G_Z \times \sigma_Z\right)^2}$$

$$= \sqrt{\begin{array}{c}\left(10^{-4} \times 10^5\right)^2 + (-2 \times 2)^2 + \left(-2.5 \times 2.083 \times 10^{-2}\right)^2 \\ + \left(6 \times 10^5 \times 2 \times 10^{-5}\right)^2\end{array}}$$

$$= 16.1246. \quad \text{(g)}$$

Per Equation (3.33) and data in Equations (f) and (g), the reliability index β is:

$$\beta = \frac{\mu_g}{\sigma_g} = \frac{40}{16.1246} = 2.48068. \quad \text{(h)}$$

Per Equation (3.34) and data in Equation (e), we have the reliability R with the use of MATLAB command:

$$R = \Phi(\beta) = \Phi(2.48068) = normcdf(2.48068, 0, 1) = 0.9934.$$

Based on the data in Equations (e) and (h), the reliability index β of the same component are different when different limit state functions are used even though both limit state functions specify the same event. The reason for this difference is because the FOSM method for a non-linear limit state function provides an approximate result. ∎

3.6   THE HASOFER–LIND (H-L) METHOD

When all variables are statistically independent, normally distributed random variables, the Hasofer–Lind (H-L) method [1] provides a more accurate and unique result of the reliability of a component with a nonlinear limit state function. The main difference between the H-L method and the FOSM method is that the H-L method will linearize the non-limit state function at the design point. The design point is a point on the surface of the limit state function: $g(X_1, X_2, \ldots, X_n) = 0$, instead of the mean-value point. Since the design point is generally not known in advance, the H-L method is an iterative process to calculate the reliability of a component with a convergence condition.

Consider the following nonlinear limit state function, which consists of mutually independent, normally distributed random variables:

$$g(X_1, X_2, \ldots, X_n) = \begin{cases} > 0 & \text{Safe} \\ = 0 & \text{Limit state} \\ < 0 & \text{Failure,} \end{cases} \tag{3.35}$$

where X_i ($i = 1, 2, \ldots, n$) is a normal distributed random variable with corresponding a mean μ_{X_i} and a standard deviation σ_{X_i}. The following equation defines the surface of a limit state function:

$$g(X_1, X_2, \ldots, X_n) = 0. \tag{3.36}$$

The general procedure for the H-L method is explained and displayed here.

Step 1: Pick an initial design point $P^{*0}\left(X_1^{*0}, X_2^{*0}, \ldots, X_n^{*0}\right)$.

The initial design point could be any point, but it must be on the surface of the limit state function as specified by Equation (3.36). We can use the mean values for the first $n-1$ variables, as shown in Equation (3.37) and then determine the last one through Equation (3.38a):

$$X_i^{*0} = \mu_{X_i} \qquad i = 1, 2, \ldots, n-1 \tag{3.37}$$

$$g\left(X_1^{*0}, X_2^{*0}, \ldots X_{n-1}^{*0}, X_n^{*0}\right) = 0 \tag{3.38a}$$

There is only one unknown X_n^{*0} in Equation (3.38a). We can solve this unknown X_n^{*0} from Equation (3.38b). When the actual limit state function is provided, we can rearrange the second equation in Equation (3.38a) and express X_n^{*0} by using $X_1^{*0}, X_2^{*0}, \ldots$, and X_{n-1}^{*0}. Let's use the following equation to represent this:

$$X_n^{*0} = g_1\left(X_1^{*0}, X_2^{*0}, \ldots, X_{n-1}^{*0}\right). \tag{3.38b}$$

Step 2: Set $\beta = 0$.

This setting is only for the MATLAB program. This setting will make sure that there are at least two iterative loops for the iterative process.

Step 3: Calculate the initial design point in the standard normal distribution space.

In the standard normal distribution space, we convert a normal distribution X_i into a standard normal distribution Z_i through the following conversion equation:

$$Z_i = \frac{X_i - \mu_{X_i}}{\sigma_{X_i}} \quad i = 1, \ldots, n. \tag{3.39}$$

Now, the surface of the limit state function in the standard normal distribution space can be expressed as:

$$g(Z_1, Z_2, \ldots, Z_n) = 0. \tag{3.40}$$

The initial design point $P^{*0}\left(X_1^{*0}, X_2^{*0}, \ldots, X_n^{*0}\right)$ in the original normal distribution space can be expressed by $P^{*0}\left(Z_1^{*0}, Z_2^{*0}, \ldots, Z_n^{*0}\right)$ in the standard normal distribution space through Equation (3.39). $Z_i^{*0}(i = 1, \ldots, n)$ can be calculated per Equation (3.41):

$$Z_i^{*0} = \frac{X_i^{*0} - \mu_{X_i}}{\sigma_{X_i}} \quad i = 1, \ldots, n. \tag{3.41}$$

Step 4: Calculate the reliability index β^{*0} at the design point $P^{*0}\left(Z_1^{*0}, Z_2^{*0}, \ldots, Z_n^{*0}\right)$.

In the H-L method, the limit state function $g(Z_1, Z_2, \ldots, Z_n)$ is linearized at the design point $P^*\left(Z_1^*, Z_2^*, \ldots, Z_n^*\right)$ through the Taylor Series. The Taylor Series coefficient, in this case, will be:

$$G_i|_{P*} = \frac{\partial g(Z_1, Z_2, \ldots, Z_n)}{\partial Z_i}\bigg|_{atP^*\left(Z_1^*, Z_2^*, \ldots, Z_n^*\right)} \quad i = 1, 2, \ldots, n, \tag{3.42}$$

where $G_i|_{P*}$ means the Taylor Series coefficient for the variable Z_i at the design point $P^*\left(Z_1^*, Z_2^*, \ldots, Z_n^*\right)$. According to the conversion Equation (3.39), we have:

$$\frac{\partial X_i}{\partial Z_i} = \sigma_{X_i}. \tag{3.43}$$

Equation (3.42) can be rewritten as:

$$\begin{aligned}
G_i|_{P*} &= \frac{\partial g(Z_1, Z_2, \ldots, Z_n)}{\partial Z_i}\bigg|_{atP^*\left(Z_1^*, Z_2^*, \ldots, Z_n^*\right)} = \frac{\partial g(Z_1, Z_2, \ldots, Z_n)}{\partial X_i}\frac{\partial X_i}{\partial Z_i}\bigg|_{atP^*\left(Z_1^*, Z_2^*, \ldots, Z_n^*\right)} \\
&= \sigma_{X_i}\frac{\partial g(X_1, X_2, \ldots, X_n)}{\partial X_i}\bigg|_{atP^*\left(X_1^*, X_2^*, \ldots, X_n^*\right)}.
\end{aligned} \tag{3.44}$$

Typically, we like to use Equation (3.44) to calculate the Taylor Series coefficient.

Now, use the FOSM method to the limit state function $g(Z_1, Z_2, \ldots, Z_n)$ at the initial design point $P^{*0}(Z_1^{*0}, Z_2^{*0}, \ldots, Z_n^{*0})$ to calculate the Taylor series coefficient $G_i|_{P*0}$ per Equation (3.45) and then calculate the reliability index β^{*0} per Equation (3.46):

$$G_i|_{P*0} = \sigma_{X_i} \left.\frac{\partial g(X_1, X_2, \ldots, X_n)}{\partial X_i}\right|_{atP^{*0}(X_1^{*0}, X_2^{*0}, \ldots, X_n^{*0})} \qquad i = 1, 2, \ldots, n \qquad (3.45)$$

$$\beta^{*0} = \frac{\sum_{i=1}^{n}\left(-Z_i^{*0} G_i|_{P*0}\right)}{\sqrt{\sum_{i=1}^{n}\left(G_i|_{P*0}\right)^2}} \qquad (3.46)$$

Step 5: Determine the new design point $P^{*1}(Z_1^{*1}, Z_2^{*1}, \ldots, Z_n^{*1})$.

The recurrence equation for the iterative process in the H-L method is the following equation:

$$Z_i^{*1} = \frac{-G_i|_{P*0}}{\sqrt{\sum_{i=1}^{n}\left(G_i|_{P*0}\right)^2}}\beta^{*0} \qquad i = 1, 2, \ldots, n-1. \qquad (3.47)$$

Since the new design point $P^{*1}(Z_1^{*1}, Z_2^{*1}, \ldots, Z_n^{*1})$ is on the surface of the limit state function $g(Z_1, Z_2, \ldots, Z_n) = 0$, the Z_n^{*1} will be obtained from the surface of the limit state function. Since we typically still use the limit state function $g(X_1, X_2, \ldots, X_n) = 0$ to conduct the calculation, we will use the following equations to get the Z_n^{*1}.

We can use the conversion Equation (3.39) to get the first $n-1$ values of the new design point $P^{*1}(X_1^{*1}, X_2^{*1}, \ldots X_{n-1}^{*1}, X_n^{*1})$ per Equation (3.48):

$$X_i^{*1} = \mu_{X_i} + \sigma_{X_i} \times Z_i^{*1}. \qquad (3.48)$$

The value X_n^{*1} is obtained per Equation (3.38b), that is,

$$X_n^{*1} = g_1\left(X_1^{*1}, X_2^{*1}, \ldots, X_{n-1}^{*1}\right). \qquad (3.49)$$

When the X_n^{*1} is obtained per Equation (3.49), Z_n^{*1} can be calculated through the conversion Equation (3.39):

$$Z_n^{*1} = \frac{X_n^{*0} - \mu_{X_n}}{\sigma_{X_n}}. \qquad (3.50)$$

Now we have the new design point $P^{*1}(X_1^{*1}, X_2^{*1}, \ldots, X_n^{*1})$ in the original normal distribution space and the same design point $P^{*1}(Z_1^{*1}, Z_2^{*1}, \ldots, Z_n^{*1})$ in the standard normal distribution space.

Step 6: Check convergence condition.

The convergence equation for this iterative process will be the difference $|\Delta\beta^*|$ between the current reliability index and the previous reliability index. Since β is a reliability index, the following convergence condition will provide an accurate estimation of the reliability:

$$|\Delta\beta^*| \leq 0.0001. \qquad (3.51)$$

If the convergence condition is satisfied, the reliability of the component will be:

$$R = P\left[g\left(X_1, X_2, \ldots, X_n\right) > 0\right] = \Phi\left(\beta^{*0}\right). \tag{3.52}$$

If the convergence condition is not satisfied, we use this new design point $P^{*1}\left(Z_1^{*1}, Z_2^{*1}, \ldots, Z_n^{*1}\right)$ to replace the previous design point $P^{*0}\left(Z_1^{*0}, Z_2^{*0}, \ldots, Z_n^{*0}\right)$, that is,

$$\begin{aligned} X_i^{*0} &= X_1^{*1} \\ Z_i^{*0} &= Z_i^{*0} \qquad i = 1, \ldots, n \\ \beta &= \beta^{*0} \end{aligned} \tag{3.53}$$

Then go to Step 4 for a new iterative process again until the convergence condition is satisfied.

Since the H-L method is an iterative process, we should use the program for calculation. The program flowchart for the H-L method is shown in Figure 3.6.

We will use the H-L method to calculate the reliability of component in Example 3.10.

Example 3.11 (Redo Example 3.10 by the H-L method)

A simple support beam is under a concentrated loading at the middle of the beam, as shown in Figure 3.7. The yield strength S_y, the concentrated load P, the beam span L and the section modulus Z of the beam are all normal distributed random variables. Their distribution parameters are: $\mu_{S_y} = 6 \times 10^5$ (kN/m²), $\sigma_{S_y} = 10^5$ (kN/m²); $\mu_P = 10$ (kN/), $\sigma_P = 2$ (kN); $\mu_L = 8$ (m), $\sigma_L = 2.083 \times 10^{-2}$ (m); and $\mu_Z = 10^{-4}$ (m³), $\sigma_{S_y} = 2 \times 10^{-5}$ (m³). Use the H-L method to calculate the reliability of the beam based on the following two different limit state function.

(1) The yield strength is the component strength index. So, the limit state function is:

$$g\left(S_y, Z, P, L,\right) = S_y - \frac{PL}{4Z} = \begin{cases} > 0 & \text{Safe} \\ = 0 & \text{Limit state} \\ < 0 & \text{Failure.} \end{cases} \tag{1}$$

(2) The allowable bending moment is the component strength index. So, the limit state function is:

$$g\left(S_y, Z, P, L\right) = S_y Z - \frac{PL}{4} = \begin{cases} > 0 & \text{Safe} \\ = 0 & \text{Limit state} \\ < 0 & \text{Failure.} \end{cases} \tag{2}$$

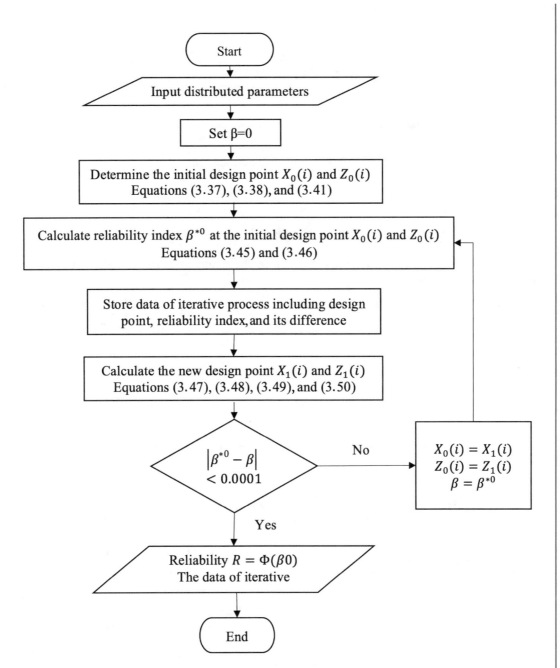

Figure 3.6: The program flowchart for the H-L method.

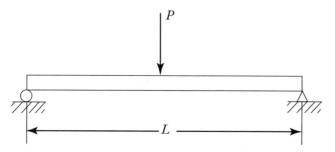

Figure 3.7: A simple support beam with a concentrated loading.

Solution:

(1) The reliability of the beam by using the limit state function: $g\left(S_y, Z, P, L\right) = S_y - \frac{PL}{4Z}$.

Per this limit state function, we can establish the following equations before we compile the program.

Per the surface of the limit state function: $g\left(S_y, Z, P, L\right) = S_y - \frac{PL}{4Z} = 0$, we can get the explicit form of Equation (3.49) for this example to determine the value of the last variable at the design point. It is:

$$L = g_1\left(S_y, Z, P\right) = \frac{S_y \times 4Z}{P}.$$ (a)

The Taylor series coefficients in this case are:

$$G_{S_y} = \sigma_{S_y} \left. \frac{\partial g\left(S_y, Z, P, L\right)}{\partial S_y} \right|_{at P*} = \sigma_{S_y} \times 1 \big|_{at P*} = \sigma_{S_y}$$ (b)

$$G_Z = \sigma_Z \left. \frac{\partial g\left(S_y, Z, P, L\right)}{\partial Z} \right|_{at P*} = \sigma_Z \times \left. \frac{PL}{4Z^2} \right|_{at P*} = \sigma_Z \frac{P^*L^*}{4(Z^*)^2}$$ (c)

$$G_P = \sigma_P \left. \frac{\partial g\left(S_y, Z, P, L\right)}{\partial P} \right|_{at P*} = \sigma_P \times \left. \left(-\frac{L}{4Z}\right) \right|_{at P*} = -\sigma_P \frac{L^*}{4Z^*}$$ (d)

$$G_L = \sigma_L \left. \frac{\partial g\left(S_y, Z, P, L\right)}{\partial L} \right|_{at P*} = \sigma_L \times \left. \left(-\frac{P}{4Z}\right) \right|_{at P*} = -\sigma_P \frac{P^*}{4Z^*}.$$ (e)

We can follow the program flowchart in Figure 3.6 to compile the MATLAB program for this example. This program is listed as "A.1: The H–L method for Example 3.11" in Appendix A.

The data for the iterative process is shown in Table 3.1. The first column is the number of iterative processes. The second to the fifth columns are the values of the design point. The

Table 3.1: The iterative results for the limit state function (1) in Example 3.11

#	S_y^*	Z^*	P^*	L^*	β^*	$|\Delta \beta^*| < 0.0001$
1	600000	0.0001	10	24	2.030685	
2	496907.9	7.53E-05	12.47421	11.99154	2.885681	0.854996
3	443008.1	5.85E-05	12.5015	8.297361	2.953815	0.068134
4	448340.9	5.41E-05	12.1497	7.983943	2.947101	0.006714
5	457762.6	5.28E-05	12.0995	7.996524	2.945587	0.001514
6	462281	5.23E-05	12.08414	7.999504	2.94526	0.000327
7	464338.3	5.2E-05	12.07591	8.000162	2.945191	6.93E-05

sixth column is the reliability index β, and the last column is the convergence condition. Per this iterative results, the reliability index β and corresponding reliability R of this component are :

$$\beta = 2.9452 \qquad R = \Phi(\beta) = \Phi(2.9452) = 0.9839.$$

(2) The reliability of the beam by using the limit state function: $g(S_y, Z, P, L) = S_y Z - \dfrac{PL}{4}$.

Per the surface of the limit state function: $g(S_y, Z, P, L) = S_y Z - \frac{PL}{4} = 0$, we can get the explicit form of Equation (3.49) for this example to determine the value of the last variable at the design point. It is:

$$L = g_1(S_y, Z, P) = \frac{S_y \times 4Z}{P}. \tag{f}$$

The Taylor series coefficients in this case are:

$$G_{S_y} = \sigma_{S_y} \left. \frac{\partial g(S_y, Z, P, L)}{\partial S_y} \right|_{atP*} = \sigma_{S_y} \times Z|_{atP*} = \sigma_{S_y} Z^* \tag{g}$$

$$G_Z = \sigma_Z \left. \frac{\partial g(S_y, Z, P, L)}{\partial Z} \right|_{atP*} = \sigma_Z \times S_y|_{atP*} = \sigma_Z S_Y^* \tag{h}$$

$$G_P = \sigma_P \left. \frac{\partial g(S_y, Z, P, L)}{\partial P} \right|_{atP*} = \sigma_P \times \left(-\frac{L}{4} \right) \bigg|_{atP*} = -\sigma_P \frac{L^*}{4} \tag{i}$$

$$G_L = \sigma_L \left. \frac{\partial g(S_y, Z, P, L)}{\partial L} \right|_{atP*} = \sigma_L \times \left(-\frac{P}{4} \right) \bigg|_{atP*} = -\sigma_P \frac{P^*}{4} \tag{j}$$

We can follow the H-L method's procedure and the program flowchart in Figure 3.6 to compile a MATLAB program which is similar to the program for the limit state function (1).

The iterative results are listed in Table 3.2. From the table, the reliability index β and corresponding reliability R of this component are:

$$\beta = 2.9452 \quad R = \Phi(\beta) = \Phi(2.9452) = 0.9839.$$

Table 3.2: The iterative results for the limit state function (2) in Example 3.11

#	S_y^*	Z^*	P^*	L^*	β^*	$|\Delta \beta^*| < 0.0001$
1	600000	0.0001	10	24	2.030685	
2	496907.9	7.53E-05	12.47421	11.99154	2.885681	0.854996
3	443008.1	5.85E-05	12.5015	8.297361	2.953815	0.068134
4	448340.9	5.41E-05	12.1497	7.983943	2.947101	0.006714
5	457762.6	5.28E-05	12.0995	7.996524	2.945587	0.001514
6	462281	5.23E-05	12.08414	7.999504	2.94526	0.000327
7	464338.3	5.2E-05	12.07591	8.000162	2.945191	6.93E-05

When the H-L method is used, the reliability of the component in two different versions of the limit state function of the same problem will be the same. ∎

3.7   THE RACKWITZ AND FIESSLER (R-F) METHOD

When a limit state function of a component contains at least one non-normal distributed random variables such as log-normal distribution or Weibull distribution, we need to use the R-F (Rackwitz and Fiessler) method [3] to calculate the reliability of a component. The R-F method is a modified H-L method. The R-F method still linearizes the limit state function at the design point, which is the point on the surface of the limit state function. The key difference between the R-F method and the H-L method is that any non-normally distributed random variable at the design point will be first converted into an equivalent normally distributed random variable, and then the H-L method is applied for calculating the reliability index.

In the R-F method, a non-normally distributed variable is converted into an equivalent normal distribution variable at the design point. Let $f_X(x)$ and $F_X(x)$ represent the PDF and the CDF of a non-normally distributed random variable X. Let $\mu_{X_{eq}}$ and $\sigma_{X_{eq}}$ represent the equivalent mean and the equivalent standard deviation of the equivalent normally distributed variable at the design point X^*. Two conditions for calculating the equivalent mean and the equivalent standard deviation of the equivalent normal distribution at the design point are as follows.

(1) The PDF of a non-normal distribution variable at the design point will be equal to the PDF of its equivalent normal distribution at the design point:

$$f_X(x^*) = \frac{1}{\sqrt{2\pi}\sigma_{X_{eq}}} e^{-\frac{(X^* - \mu_{X_{eq}})^2}{2\sigma_{X_{eq}}^2}} = \frac{1}{\sigma_{X_{eq}}} \phi\left(\frac{X^* - \mu_{X_{eq}}}{\sigma_{X_{eq}}}\right), \tag{3.54}$$

where $\phi(\cdot)$ is the PDF of the standard normally distributed random variable.

(2) The CDF of a non-normal distribution variable at the design point will be equal to the CDF of its equivalent normal distribution at the design point:

$$F_X(X^*) = \int_{-\infty}^{X^*} \frac{1}{\sqrt{2\pi}\sigma_{X_{eq}}} e^{-\frac{(x^* - \mu_{X_{eq}})^2}{2\sigma_{0X_{eq}}^2}} dx = \Phi\left(\frac{X^* - \mu_{X_{eq}}}{\sigma_{X_{eq}}}\right), \tag{3.55}$$

where $\Phi(\cdot)$ is the CDF of the standard normally distributed random variable.

For the convenient calculation, we can introduce z_X^*, which is the value of the standardized equivalent normal distribution at the design point x^*:

$$z_X^* = \frac{x^* - \mu_{X_{eq}}}{\sigma_{X_{eq}}}. \tag{3.56}$$

We can rewrite the Equations (3.54) and (3.55) as:

$$f_X(x^*) = \frac{1}{\sigma_{X_{eq}}} \phi(z_X^*) \tag{3.57}$$

$$F_X(X^*) = \Phi(z_X^*) \tag{3.58}$$

From Equation (3.58) we can get z_X^*:

$$z_X^* = \Phi^{-1}[F_X(X^*)], \tag{3.59}$$

where $\Phi^{-1}(\cdot)$ is the inverse of the CDF of the standard normal distribution.

In Excel, the function for $\Phi^{-1}(\cdot)$ is:

$$z_X^* = \Phi^{-1}[F_X(X^*)] = NORM.S.INV(F_X(X^*)). \tag{3.60}$$

In the MATLAB program, the command for $\Phi^{-1}(\cdot)$ is:

$$z_X^* = \Phi^{-1}[F_X(X^*)] = norminv(F_X(X^*)). \tag{3.61}$$

Since $F_X(X^*)$ of the non-normally distribution at the design point X^* is known, z_X^* can be determined by Equations (3.60) or (3.61). Then, from Equation (3.57), we can calculate the equivalent standard deviation $\sigma_{X_{eq}}$:

$$\sigma_{X_{eq}} = \frac{1}{f_X(X^*)} \phi(z_X^*).$$ (3.62)

After the equivalent standard deviation $\sigma_{X_{eq}}$ is known, we can calculate the equivalent mean $\mu_{X_{eq}}$ from Equation (3.56):

$$\mu_{X_{eq}} = x^* - z_X^* \times \sigma_{X_{eq}}.$$ (3.63)

We can use Equations (3.59)–(3.63) to calculate the equivalent mean $\mu_{X_{eq}}$ and the equivalent standard deviation $\sigma_{X_{eq}}$ of the equivalent normal distribution of a non-normal distribution at the design point X^*.

Example 3.12

A random variable X follows a log-normal distribution with the distribution parameter $\mu_{\ln X} = 7.81$ and the $\sigma_{\ln X} = 0.192$.

(a) Use Excel to determine the equivalent mean and the equivalent standard deviation of the equivalent normal distribution at the design point $X^* = 2600$.

(b) Use MATLAB to determine the equivalent mean and the equivalent standard deviation of the equivalent normal distribution at the design point $X^* = 2600$.

Solution:

(a) Use Excel to determine $\mu_{X_{eq}}$ and $\sigma_{X_{eq}}$ at the design point $X^* = 2600$.

The PDF is $f_X(X^*)$ and the CDF is $F_X(X^*)$ of the log-normal distribution at the design point $X^* = 2600$ with the distribution parameters $\mu_{\ln X} = 7.81$ and the $\sigma_{\ln X} = 0.192$ are:

$$f_X(2600) = LOGNORM.DIST(2600, 7.81, 0.192 FALSE) = 7.68993 \times 10^{-4}$$ (a)

$$F_X(2600) = LOGNORM.DIST(2600, 7.81, 0.192, TRUE) = 0.609275,$$ (b)

where $LOGNORM.DIST(x, \mu_{\ln X}, \sigma_{\ln X}, FALSE \, or \, TRUE)$ is the PDF with "FALSE" and the CDF with "TRUE" of Excel function for a log-normal distribution.

From Equations (3.60) and (b), we have:

$$z_X^* = \Phi^{-1}[F_X(x^*)] = NORM.S.INV(0.609275) = 0.277430,$$ (c)

where $NORM.S.INV(z)$ is the Excel function for the inverse of the standard normal distribution with a probability z.

From Equations (3.62), (a), and (c), we can have the equivalent standard deviation $\sigma_{X_{eq}}$:

$$\sigma_{X_{eq}} = \frac{1}{f_X\,(x^*)}\phi\,(z_X^*) = \frac{NORM.S.DIST\,(0.277430)}{7.68993 \times 10^{-4}} = \frac{0.383881}{7.68993 \times 10^{-4}} = 499.2. \quad \text{(d)}$$

From Equations (3.63), (c), and (d), we can have the equivalent mean $\mu_{X_{eq}}$:

$$\mu_{X_{eq}} = X^* - z_X^* \times \sigma_{X_{eq}} = 2600 - 0.277430 \times 499.20 = 2461.5. \quad \text{(e)}$$

(b) Use the MATLAB program to determine $\mu_{X_{eq}}$ and $\sigma_{X_{eq}}$ at the design point $X^* = 2600$.

From Equations (3.57), (3.60), and (3.61), we can compile a short program as listed here.

```
% Calculate an equivalent mean and standard deviation
% Input the lognormal distribution parameter
mln=7.81;        % Log mean
sln=0.192;       % Log standard deviation
xs=2600;         % Design point x*
fx=lognpdf (xs, mln, sln); % PDF of log-normal distribution
Fx=logncdf (xs, mln, sln); % CDF of log-normal distribution
zs=norminv (Fx);     % The inverse value of the normal
                     % distribution per Equation (3.61)
seq=normpdf (zs)/fx; % Equivalent standard deviation per
                     % Equation (3.64)
meq=xs-seq*zs;       % Equivalent mean, Equation (3.65)
display ('Equivalent mean')
meq
display ('Equivalent standard deviation')
seq
```

MATLAB gives the following results:
     Equivalent mean = 2461.5.
     Equivalent standard deviation = 499.2.   ∎

The R-F method is an iterative process. The procedure is very similar to the H-L method. However, at the beginning of each iterative process, any non-normally distributed random variable will be converted into an equivalent normal distribution with the equivalent mean and the equivalent standard deviation. Following is the general procedure for the R-F method.

Step 1: Calculate the mean for non-normal distributed random variables.

For a clear description of the R-F method, we can rearrange the limit state function, as shown in Equation (3.64). In Equation (3.64), the first r random variables are non-normally distributed random variables, and the rest $(n - r)$ random variables are normally distributed random variables:

$$g(X_1, \ldots, X_r, X_{r+1}, \ldots, X_n) = \begin{cases} > 0 & \text{Safe} \\ = 0 & \text{Limit state} \\ < 0 & \text{Failure.} \end{cases} \tag{3.64}$$

The surface of this limit state function is

$$g(X_1, \ldots, X_r, X_{r+1}, \ldots, X_n) = 0. \tag{3.65}$$

For non-normally distributed random variable, we can calculate their means based on their type of distributions, which has been discussed in Chapter 2. The means of some typical distribution are listed here.

For a uniform distribution X with distribution parameters a and b, the mean is

$$\mu_X = \frac{a + b}{2}. \tag{3.66}$$

For a log-normal distribution X with the distribution parameter $\mu_{\ln x}$ and $\sigma_{\ln x}$, the mean is

$$\mu_X = Exp\left(\mu_{\ln x} + \frac{\sigma_{\ln x}^2}{2}\right). \tag{3.67}$$

For a two-parameter Weibull distribution X with the scale parameter η and the shape parameter β, the mean is

$$\mu_X = \eta\Gamma\left(\frac{1}{\beta} + 1\right). \tag{3.68}$$

For an exponential distribution X with the distribution parameter λ, the mean is

$$\mu_X = \frac{1}{\lambda}. \tag{3.69}$$

Step 2: Pick an initial design point $P^{*0}\left(X_1^{*0}, X_2^{*0}, \ldots, X_n^{*0}\right)$.

The initial design point could be any point. The simple choice is to use the means of every random variable as the design point. Since the design point must be on the surface of the limit state function as specified by Equation (3.65), we can use the mean values for the first $n - 1$ variables, and then determine the last one through Equation (3.71a). Since there is only one unknown X_n^{*0} in Equation (3.71a), we can solve this unknown X_n^{*0}:

$$X_i^{*0} = \mu_{X_i} \qquad i = 1, 2, \ldots, n - 1 \tag{3.70}$$

$$g\left(X_1^{*0}, X_2^{*0}, \ldots X_{n-1}^{*0}, X_n^{*0}\right) = 0. \tag{3.71a}$$

When the actual limit state function is provided, we can rearrange the above equation and express X_n^{*0} by using $X_1^{*0}, X_2^{*0}, \ldots$, and X_{n-1}^{*0}. Let's use the following equation to represent this:

$$X_n^{*0} = g_1 \left(X_1^{*0}, X_2^{*0}, \ldots, X_{n-1}^{*0} \right). \tag{3.71b}$$

Now, we have the initial design point $P^{*0} \left(X_1^{*0}, X_2^{*0}, \ldots X_{n-1}^{*0}, X_n^{*0} \right)$.

Step 3: Set $\beta = 0$.

This setting is only for MATLAB. This setting will make sure that the iterative process will have at least two iterative loops.

Step 4: The mean and standard deviation at the design point $P^{*0} \left(X_1^{*0}, X_2^{*0}, \ldots X_{n-1}^{*0}, X_n^{*0} \right)$.

For non-normally distributed random variables, we convert them into equivalent normal distributed random variables per Equations (3.61), (3.62), and (3.63).

For the first r random variables in the limit state function described in Equation (3.64), we have:

$$z_{X_i}^{*0} = \Phi^{-1} \left[F_{X_i} \left(X_i^{*0} \right) \right] = norminv \left(F_{X_i} \left(X_i^{*0} \right) \right)$$

$$\sigma_{X_i eq} = \frac{1}{f_{X_i} \left(X_i^{*0} \right)} \phi \left(z_{X_i}^{*0} \right) \qquad i = 1, 2, \ldots, r \tag{3.72}$$

$$\mu_{X_i eq} = x_i^{*0} - z_{X_i}^{*0} \times \sigma_{X_i eq},$$

where x_i^{*0} is the value of the non-normally distributed random variable X_i at the design point $P^{*0} \left(X_1^{*0}, X_2^{*0}, \ldots X_{n-1}^{*0}, X_n^{*0} \right)$. $f_{X_i} \left(x_i^{*0} \right)$ and $F_{X_i} \left(x_i^{*0} \right)$ are the PDF and CDF of the non-normally distributed random variable X_i at the design point X_i^{*0}. $\mu_{X_i eq}$ and $\sigma_{X_i eq}$ are the equivalent mean and the equivalent standard deviation of the equivalent normally distributed random variable at the design point x_i^{*0}.

Now every random variable in the limit state function in Equation (3.64) at the design point P^{*0} are normally distributed random variables. The mean and standard deviation of these normally distributed random variables are

$$\mu_{X_i} = \begin{cases} \mu_{X_i eq} & i = 1, 2, \ldots, r \\ \mu_{X_i} & i = r + 1, \ldots n \end{cases} \tag{3.73}$$

$$\sigma_{X_i} = \begin{cases} \sigma_{X_i eq} & i = 1, 2, \ldots, r \\ \sigma_{X_i} & i = r + 1, \ldots n. \end{cases} \tag{3.74}$$

Step 5: Calculate the initial design point P^{*0} in the standard normal distribution space.

In the standard normal distribution space, we convert a normal distribution X_i into a standard normal distribution Z_i through the following conversion equation:

$$Z_i = \frac{X_i - \mu_{X_i}}{\sigma_{X_i}} \qquad i = 1, \ldots, n. \tag{3.75}$$

Now, the limit state function in the standard normal distribution space can be expressed as:

$$g\left(Z_1, Z_2, \ldots, Z_n\right) = \begin{cases} > 0 & \text{Safe} \\ = 0 & \text{Limit state} \\ < 0 & \text{Failure.} \end{cases} \qquad (3.76)$$

The initial design point $P^{*0}\left(X_1^{*0}, \ldots, X_n^{*0}\right)$ can be expressed as $P^{*0}\left(Z_1^{*0}, Z_2^{*0}, \ldots, Z_n^{*0}\right)$ in the standard normal distribution space through Equation (3.77):

$$Z_i^{*0} = \frac{X_i^{*0} - \mu_{X_i}}{\sigma_{X_i}} \qquad i = 1, \ldots, n. \qquad (3.77)$$

Step 6: Calculate the reliability index β^{*0} at the design point $P^{*0}\left(Z_1^{*0}, Z_2^{*0}, \ldots, Z_n^{*0}\right)$.

This step is the same as Step 4 for the H-L method. We can use the FOSM method to linearize the limit state function as shown in Equation (3.76) at the initial design point $P^{*0}\left(Z_1^{*0}, Z_2^{*0}, \ldots, Z_n^{*0}\right)$. Then, per Equation (3.78) we can calculate the Taylor series coefficients, and per Equation (3.79) we can calculate the reliability index β^{*0}:

$$G_i|_{P*0} = \sigma_{X_i} \left. \frac{\partial g\left(X_1, X_2, \ldots, X_n\right)}{\partial X_i} \right|_{at P^{*0}\left(X_1^{*0}, X_2^{*0}, \ldots, X_n^{*0}\right)} \qquad i = 1, 2, \ldots, n \qquad (3.78)$$

$$\beta^{*0} = \frac{\sum_{i=1}^n \left(-Z_i^{*0} G_i|_{P*0}\right)}{\sqrt{\sum_{i=1}^n \left(G_i|_{P*0}\right)^2}}. \qquad (3.79)$$

Step 7: Determine the new design point $P^{*1}\left(Z_1^{*1}, Z_2^{*1}, \ldots, Z_n^{*1}\right)$ for the iterative process.

The recurrence equations for the iterative process are the following equations:

$$Z_i^{*1} = \frac{-G_i|_{P*0}}{\sqrt{\sum_{i=1}^n \left(G_i|_{P*0}\right)^2}} \beta^{*0} \qquad i = 1, 2, \ldots, n - 1. \qquad (3.80)$$

Because the new design point $P^{*1}\left(Z_1^{*1}, Z_2^{*1}, \ldots, Z_n^{*1}\right)$ is on the surface of the limit state function $g\left(Z_1, Z_2, \ldots, Z_n\right) = 0$, the Z_n^{*1} will be obtained from the surface of the limit state function. Since we typically still use the limit state function $g\left(X_1, X_2, \ldots, X_n\right) = 0$ to conduct the calculation, we will use the following equations to get the Z_n^{*1}.

We can use the conversion Equation (3.75) to get the first $n - 1$ values of the new design point $P^{*1}\left(X_1^{*1}, X_2^{*1}, \ldots X_{n-1}^{*1}, X_n^{*1}\right)$, that is:

$$X_{1i}^{*1} = \mu_{X_i} + \sigma_{X_i} \times Z_i^{*1}. \qquad (3.81)$$

Per the surface of the limit state function Equation (3.65), we can rearrange the limit state function in such a way that X_n is expressed as the function of $X_1, X_2, \ldots$, and X_{n-1}, that is, $X_n = g_1(X_1, X_2, \ldots, X_{n-1})$. The value X_n^{*1} is obtained per Equation (3.82):

$$X_n^{*1} = g_1\left(X_1^{*1}, X_2^{*1}, \ldots, X_{n-1}^{*1}\right). \tag{3.82}$$

After the X_n^{*1} is obtained from Equation (3.82), Z_n^{*1} can be calculated through the conversion Equation (3.83):

$$Z_n^{*1} = \frac{X_n^{*0} - \mu_{X_n}}{\sigma_{X_n}}. \tag{3.83}$$

Now we have the new design point $P^{*1}\left(X_1^{*1}, X_2^{*1}, \ldots, X_n^{*1}\right)$ in original normal distribution space and the same design point $P^{*1}\left(Z_1^{*1}, Z_2^{*1}, \ldots, Z_n^{*1}\right)$ in the standard normal distribution space.

Step 8: Check convergence condition.

The convergence equation for this iterative process will be the difference $\left|\Delta\beta^{*0}\right|$ between the current reliability index and the previous reliability index. Since β is a reliability index, the following convergence condition will provide an accurate estimation of the reliability:

$$\left|\Delta\beta^{*0}\right| \leq 0.0001. \tag{3.84}$$

If the convergence condition is satisfied, the reliability of the component will be:

$$R = P[g(X_1, X_2, \ldots, X_n) > 0] = \Phi(\beta^{*0}). \tag{3.85}$$

If the convergence condition is not satisfied, we use this new design point $P^{*1}\left(X_1^{*1}, X_2^{*1}, \ldots, X_n^{*1}\right)$ to replace the previous design point $P^{*0}\left(X_1^{*0}, X_2^{*0}, \ldots, X_n^{*0}\right)$, that is,

$$\begin{aligned} X_i^{*0} &= X_i^{*1} \qquad i = 1, \ldots, n \\ \beta &= \beta^{*0}. \end{aligned} \tag{3.86}$$

Then, we go to Step 4 for a new iterative process again until the convergence condition is satisfied.

The program flowchart for the R-F method is shown in Figure 3.8.

Example 3.13

The shear yield strength S_{sy} (ksi) of a shaft follows a normal distribution with a mean $\mu_{S_{sy}} = 31$ (ksi) and a standard deviation $\sigma_{S_{sy}} = 2.4$ (ksi). The diameter d (inch) of the solid shaft follows a normal distribution with a mean $2.125''$ and the standard deviation $0.002''$. The torque applied on the shaft T (klb.in) follows a two-parameter Weibull distribution with the scale parameter $\eta = 34$ and the shape parameter $\beta = 3$. Use the allowable torque to build the limit state function,

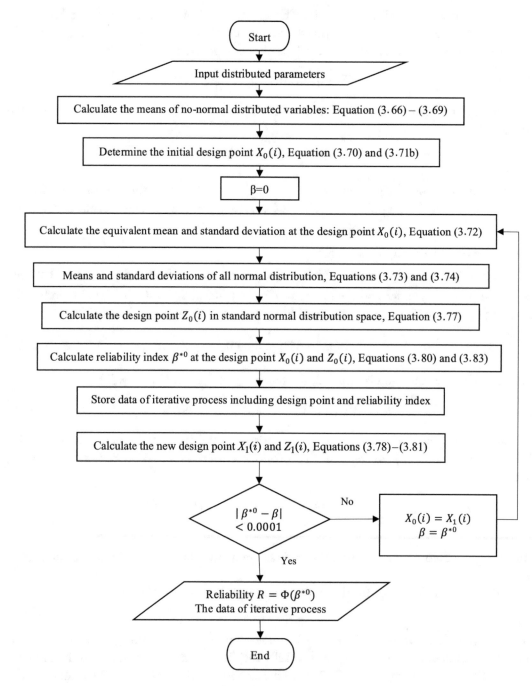

Figure 3.8: The program flowchart for the R-F method.

which is equal to the multiplication of the shear yield strength with the section modulus of the shaft. Then determine the reliability of this shaft.

Solution:

(1) Use the allowable torque to build the limit state function.

If we use the yield strength as the component strength index, the limit state function of this example will be:

$$g\left(T, d, S_{sy}\right) = S_{sy} - \frac{T}{\left(\frac{\pi}{16}\right) d^3} = \begin{cases} > 0 & \text{Safe} \\ = 0 & \text{Limit state} \\ < 0 & \text{Failure.} \end{cases} \quad \text{(a)}$$

In this example, the allowable torque is the component strength index. Therefore, we can have the following limit state function:

$$g\left(T, d, S_{sy}\right) = S_{sy}\left(\frac{\pi}{16}\right) d^3 - T = \begin{cases} > 0 & \text{Safe} \\ = 0 & \text{Limit state} \\ < 0 & \text{Failure.} \end{cases} \quad \text{(b)}$$

Both Equations (a) and (b) are limit state functions of this example. When the R-F method is used, the results will be the same by using any one of these two limit state functions. In this example, we will use the limit state function described in Equation (b) to calculate the reliability.

We can establish the following equations for the preparation of compiling the MATLAB program.

The mean for the Weibull distribution per Equation (3.68), we have:

$$\mu_X = \eta\Gamma\left(\frac{1}{\beta} + 1\right). \quad \text{(c)}$$

For this limit state function, the explicit form of Equation (3.71b) for determining the value of the last variable at the design point is

$$S_{sy} = g_1\left(T, d\right) = \frac{16T}{\pi d^3}. \quad \text{(d)}$$

In this example, we have three random variables. Two random variables d and S_{sy} are normal distributions. The torque T is a Weibull distribution. Per Equation (3.72), the equivalent mean $\mu_{T_{eq}}$ and the equivalent standard deviation $\sigma_{T_{eq}}$ of the torque T at the design point T^* are:

$$z_T^* = \Phi^{-1}\left[F_T\left(T^*\right)\right] = norminv\left[wblcdf(T^*, \eta, \beta)\right]$$

$$\sigma_{T_{eq}} = \frac{1}{f_T(T^*)}\phi\left(z_T^*\right) = \frac{1}{wblpdf(T^*, \eta, \beta)}normpdf\left(z_T^*\right) \quad \text{(e)}$$

$$\mu_{T_{eq}} = T^* - z_T^* \times \sigma_{T_{eq}},$$

where $wblpdf(T^*, \eta, \beta)$ and $wblcdf(T^*, \eta, \beta)$ are the commands for the PDF and the CDF of the Weibull distribution with the scale parameter η and the shape function β at the design point T^*. $norminv(z_T^*)$ is the command for the inverse of standard normal distribution at the given probability z_T^*.

Per Equation (3.78), the Taylor series coefficients in this example are:

$$G_T|_{P*} = \sigma_T \left.\frac{\partial g(T, d, S_{sy})}{\partial T}\right|_{atP*0} = \sigma_T (-1)|_{atP*0} = -\sigma_T$$

$$G_d|_{P*} = \sigma_d \left.\frac{\partial g(T, d, S_{sy})}{\partial d}\right|_{atP*0} = \sigma_d \left.\left(\frac{S_{sy} 3\pi d^2}{16}\right)\right|_{atP*0} = \sigma_d \frac{S_{sy}^* 3\pi (d^*)^2}{16} \qquad \text{(f)}$$

$$G_{S_{sy}}|_{P*} = \sigma_{S_{sy}} \left.\frac{\partial g(T, d, S_{sy})}{\partial S_{sy}}\right|_{atP*0} = \sigma_{S_{sy}} \left.\left(\frac{\pi d^3}{16}\right)\right|_{atP*0} = \sigma_{S_{sy}} \frac{\pi (d^*)^3}{16}.$$

We can follow the provided procedure or the flowchart in Figure 3.8 to compile the MATLAB program. The MATLAB program for this example is listed in Appendix A as "A.2: The R-F method for Example 3.13."

The data for the iterative process is shown in Table 3.3. The first column is the number of iterative processes. The second to the fourth columns are the values of the design points. The fifth column is the reliability index β, and the last column is the convergence condition.

From the iterative results, the reliability index β and corresponding reliability R of this component are:

$$\beta^{*0} = 2.3058, \qquad R = \Phi\left(\beta^{*0}\right) = \Phi(2.3058) = 0.9894.$$

The iterative results by the R-F method for Example 3.13. ∎

Table 3.3: Iterative results by the R_F method

#	T^*	d^*	S_y^*	β^*	$|\Delta \beta^*| < 0.0001$
1	30.3613	2.125	16.11438	2.28222	
2	54.64176	2.124941	29.00377	2.305446	0.023226
3	54.43365	2.124887	28.89548	2.305759	0.000314
4	54.44002	2.124888	28.89884	2.305756	2.85E-06

3.8   THE MONTE CARLO METHOD

A general limit state function $g(X_1, X_2, \ldots, X_n)$ of a component is the function of random variables $X_1, X_2, \ldots,$ and X_n. Therefore, it is also a random variable. The reliability of the component is the probability of the event $g(X_1, X_2, \ldots, X_n) \geq 0$. When the distributions of all random variables are given, we can use the interference method, the FOSM method, the H-L method, or the R-F method to calculate the reliability, which has been discussed in Sections 3.3–3.7. We can also use the relative frequency to estimate the reliability, which has been discussed in Section 2.4. For example, when a sample value x_i of each random variable is known, we can use these sample values $(x_1, x_2, \ldots, x_n)$ to calculate a sample value of the limit state function $g(x_1, x_2, \ldots, x_n)$, which may be larger than and equal to, or less than zero. This process can be called a trial in the virtual experiment. Per the definition of probability, we can use the relative frequency to estimate the reliability when the number of sample data of the limit state function is sufficiently big. The Monte Carlo method relies on repeated random sampling to obtain the numerical value of the limit state function $g(X_1, X_2, \ldots, X_n)$ for estimating the relative frequency.

Basic concepts and procedure for the Monte Carlo method are as follows.

Step 1: Uniformly and randomly generate one sample value for each random variable per its corresponding probabilistic distribution. Let x_i^{*j} ($i = 1, 2, \ldots, n$) be the sample data in the jth trial of the virtual experiment. Here, the subscript i in x_i^{*j} refers to the ith random variable X_i. The superscript j in x_i^{*j} refers to the jth trial. The x_i^{*j} is the sample value of the random variable X_i in the jth trial of the virtual experiment.

Step 2: Use x_i^{*j} ($i = 1, 2, \ldots, n$) in the limit state function to get a trial value of the limit state function. Per the definition of the limit state function, when the trial value $g\left(x_1^{*j}, x_2^{*j}, \ldots, x_n^{*j}\right)$ of the limit state function of the component is larger than or equal to zero, the component is safe. When the trail value: $g\left(x_1^{*j}, x_2^{*j}, \ldots, x_n^{*j}\right)$ of the limit state function of the component is less than zero, the component is a failure. We can use VT^{*j} to represent the trial result:

$$VT^{*j} = \begin{cases} 1 & \text{when } g\left(x_1^{*j}, x_2^{*j}, \ldots, x_n^{*j}\right) \geq 0 \\ 0 & \text{when } g\left(x_1^{*j}, x_2^{*j}, \ldots, x_n^{*j}\right) < 0, \end{cases} \tag{3.87}$$

where VT^{*j} is the trial result of the jth trial of the virtual experiment. The value "1" of the VT^{*j} indicates a safe status of the component. The value "0" of the VT^{*j} indicates a failure status of the component.

Step 3: Repeat Step 1 and Step 2 until enough trials N have been conducted.

Step 4: The relative frequency of the component with a safe status in total trial N will be the probability of the event $g(X_1, X_2, \ldots, X_n) \geq 0$. Therefore, the reliability of the component will be

$$R = P\left[g(X_1, X_2, \ldots, X_n) \geq 0\right] = \frac{\sum_{j=1}^{N}\left(VT^{*j}\right)}{N}. \tag{3.88}$$

The probability of the component failure F will be:

$$F = 1 - R = 1 - \frac{\sum_{j=1}^{N} \left(VT^{*j} \right)}{N}. \tag{3.89}$$

In the Monte Carlo method, the relative error between the true value of the probability of the component failure and the estimated value in Equation (3.89) will become smaller when the trail number N increases. For a 95% percent confidence level, the relationship [4, 5] between the relative error ε and the trial number N is:

$$\varepsilon = 2\sqrt{\frac{1 - F}{N \times F}}, \tag{3.90}$$

where ε is the relative error of the probability of component failure with a 95% confidence level. F is the estimated probability of the failure of the component per Equation (3.89). N is the trial number in the Monte Carlo method. For example, if the estimated reliability is 0.975 and the relative error of the probability of the component failure is 0.034, that is, 3.4%. In this case, the estimated probability of the failure is $F = 1 - R = 1 - 0.975 = 0.025$. Since the relative error for F is 0.034, the F will have the value in the range $0.025 \pm 0.025 \times 0.034 = 0.025 \pm 0.0008$. Therefore, the range of reliability R will be 0.975 ± 0.0008.

We can rearrange Equation (3.90) to express the trial number N as a function of relative percent error ε and the estimated reliability R:

$$N = \frac{4(1 - F)}{\varepsilon^2 F} = \frac{4R}{\varepsilon^2 (1 - R)}. \tag{3.91}$$

We can use Equation (3.91) to calculate the trial number N for the Monte Carlo method. After completing the calculation by the Monte Carlo method, we can use Equation (3.91) to calculate the relative error. To use Equation (3.91) to calculate the trial number N, we need to know the pre-specified relative error ε and the possible value of the reliability R. The possible value of the reliability R can be guested according to its importance level in mechanical design, as shown in Table 3.4.

If the pre-specified relative error of the probability of component failure is 0.05 with a 95% confidence level, the trial number N for the Monte Carlo method for the different level of importance of a component is listed in Table 3.5.

The Monte Carlo method needs to have a huge amount of trials for an acceptable accuracy of the estimated reliability. It is typically implemented through a computer program. This book will use MATLAB as the computer program to implement the Monte Carlo method. We can use the following command in the MATLAB software to generate a matrix $1 \times N$ of random numbers of a specified distributed random variable:

$$RX_i = random\left('name', A, B, 1, N \right), \tag{3.92}$$

Table 3.4: The levels of importance

Levels of Importance	Reliability	Notes
Critical component	0.9999	The failure of this component will cause death to the operators or huge financial loss.
Key component	0.999	The failure of this component will cause major financial loss or a long period of system shutdown.
Important component	0.99	The failure of this component might cause some financial loss or a short period of system shutdown.
General component	0.9	The failure of this component will not cause a big issue to the system.

Table 3.5: The estimation of the trial number N

Levels of Importance	Reliability	Relative Error	Trial Number N
Critical component	0.9999	0.05	15998400
Key component	0.999	0.05	1598400
Important component	0.99	0.05	158400
General component	0.9	0.05	14400

where RX_i is a matrix $1 \times N$ with N of random samplings of a distributed random variable X_i. *random* is the MATLAB command for generating a random number. *'name'* is to specify the type of distribution. A and B are the distribution parameters of the distributed random variables. "1, N" in Equation (3.92) means that the matrix for these random number will be stored as one row with N column. The *'name'*, A and B for several distributions are listed in Table 3.6.

Table 3.6: The *'name'*, A and B for several typical distributions

'name'	Distribution	Input parameter A	Input parameter B
'exp'	Exponential distribution	Mean $\mu = 1/\lambda$	/
'logn'	Log-normal distribution	Mean of logarithmic values, μ_{ln}	Standard deviation of logarithmic values, σ_{ln}
'norm'	Normal distribution	Mean μ	Standard deviation σ
'unif'	Uniform distribution	Lower endpoint a	Upper endpoint b
'wbl'	Weibull distribution	Scale parameter η	Shape parameter β

The flowchart of a MATLAB program for the Monte Carlo method is displayed in Figure 3.9.

Example 3.14

A key component is subjected to two axial loadings, as shown in Figure 3.10. The two axial loadings P_1 (kip) and P_2 (kip) are statistically independent random variables. P_1 follows a normal distribution with a mean $\mu_{P_1} = 10.2$ and a standard deviation $\sigma_{P_1} = 1.2$. P_2 follows a Weibull distribution with the scale parameter $\eta = 4.5$ and the shape parameter $\beta = 1.5$. The yield strength S_y of the component will be the component strength index and follows a normal distribution with a mean $\mu_{S_y} = 61.5$ (ksi) and a standard deviation $\sigma_{S_y} = 5.95$ (ksi). The diameter d of the round bar can be described by a normal distribution with a mean $\mu_d = 0.75''$ and the standard deviation $\sigma_d = 0.003''$. Compile the MATLAB program to implement the Monte Carlo method to estimate the reliability of this component with a 95% confidence level and calculate the relative error. Show the range of the reliability of the component with a 95% confidence level.

Solution:

In this example, the limit state function is:

$$g\left(S_y, P_1, P_2, d\right) = S_y - \frac{4P_1}{\pi d^2} - \frac{4P_2}{\pi d^2}. \tag{a}$$

Let's use 5% as the pre-specified relative error with a 95% confidence level. Because the component is a key component, the trial number N for this Monte Carlo method from Table 3.2 is:

$$N = 1{,}598{,}400. \tag{b}$$

We can compile the MATLAB program by following the procedure and the flowchart in Figure 3.9. The MATLAB program for this problem is listed in Appendix A with the name "A.3: The Monte Carlo Method for Example 3.14."

The estimated reliability of this component R is:

$$R = 0.9973. \tag{c}$$

The estimated probability of the component failure F is

$$F = 1 - R = 1 - 0.9973 = 0.0027. \tag{d}$$

The relative error of the probability of the failure is

$$\varepsilon = 0.0306. \tag{e}$$

So, the range of the probability of the component failure with a 95% confidence level will be:

$$F = 0.0027 \pm 0.0027 \times 0.0306 = 0.0027 \pm 0.00008. \tag{f}$$

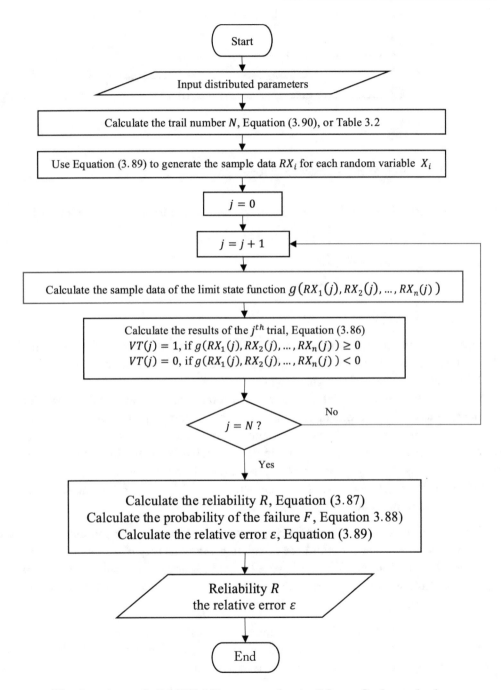

Figure 3.9: The flowchart of a MATLAB program for the Monte Carlo method.

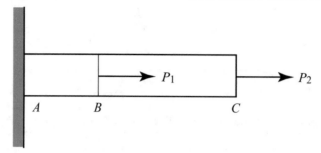

Figure 3.10: A key component is under axial loadings.

Therefore, the range of the reliability of the component with a 95% confidence level will be:

$$R = 1 - F = 0.9973 \pm 0.00008. \tag{g}$$

■

3.9 SUMMARY

The calculation of the reliability of a component is one of the key topics for reliability-based mechanical design since it replaces the traditional factor of safety as the measure of the design in reliability-based mechanical design.

We must establish the limit state function of a component to calculate the reliability of the component, which is the relationship among the reliability, component geometrical dimensions, component strength, mechanical properties, and loading. The limit state function of a component under static loadings will be discussed in detail in Chapter 4. The limit state function of a component under cyclic loadings will be discussed in detail in Chapter 1 of Volume 2.

When the limit state function of a component is established, the following is the summary of reliability calculation approaches of a component.

1. When the limit state function of a component contains only two random variables, we can use the interference method for calculating the reliability of a component, which is discussed in Section 3.3. We can have a theoretical, analytical result only for both normal distributions, both log-normal distributions, both exponential distributions, and both uniform distributions.

2. When the limit state function of a component is a linear function of all statistically independently normally distributed random variables, we can use the FOSM method to calculate the reliability, which is discussed in Section 3.5.

3. When the limit state function of a component is a nonlinear function of all statistically independently normally distributed random variables or is a linear or nonlinear function of

all statistically independent random variables with only the known means and the known standard deviations, we can still use the FOSM method to estimate the reliability of the component.

4. When the limit state function of a component is a nonlinear function of all statistically independently normally distributed random variables, we can use the H-L method to calculate the reliability of the component, which is discussed in detail in Section 3.6. The H-L method is an improved version of the FOSM method.

5. When the limit state function of a component contains one or more non-normally distributed random variables, we can use the R-F method to calculate the reliability of the component, which is discussed in detail in Section 3.7. The R-F method is an improved H-L method.

6. The Monte Carlo method can be used to calculate the reliability of a component for a limit state function with any type of random variables. The Monte Carlo method is discussed in Section 3.8 and must be implemented through a computer program because of a huge amount of trial numbers in this virtual experiment.

3.10   REFERENCES

[1] Hasofer, A. M. and Lind, N., An exact and invariant first-order reliability format, *Journal of Engineering Mechanics, ASCE*, vol. 100, no. EM1, pp. 111–121, February 1974. 124, 133

[2] Nowak, A. S. and Collins, K. R., *Reliability of Structures*, 2nd ed., CRC Press, 2013. DOI: 10.1201/b12913. 124, 130

[3] Rackwitz, R. and Fiessler, B., Structural reliability under combined random load sequences, *Computers and Structures*, vol. 9, pp. 489–494, 1978. DOI: 10.1016/0045-7949(78)90046-9. 140

[4] Shooman, M. L., *Probabilistic Reliability: An Engineering Approach*, McGraw-Hill, New York, 1968. 152

[5] Rao, S. S., *Reliability Engineering*, Pearson, 2015. 152

3.11   EXERCISES

3.1.   A double-shear pin with a diameter d is subjected to direct shearing force V. The shearing yield strength of this material is S_{sy}. The failure mode is that when the direct shear stress is more than the shearing yield strength, the pin is treated as a failure. Establish the limit state function of this pin.

3.2.   A simple support beam with a span L is subjected to a concentrated force P at the middle of the beam, as shown in Figure 3.11. The beam is a constant round bar with a diameter d. The yield strength of the beam material is S_y. It is assumed that the shear stress at the critical section induced by the sheer force is negligible. It is assumed that the beam will be treated as a failure when the maximum normal stress is more than the yield strength. Establish the limit state function of this beam.

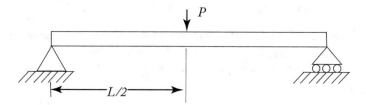

Figure 3.11: A simple support beam under a concentrated force at the middle of the beam.

3.3.   The design specification of a simple support beam, as shown in Figure 3.11 is that the maximum deflection of the beam is not allowed to be more than Δ. The Young's modulus of the beam material is E. Establish the limit state function of this beam.

3.4.   The failure mode of a pin is a static failure at which the direct shear stress τ_D is more than the yield shearing strength S_{sy}. The yield shearing strength S_{sy} and the direct shear stress τ_D are uniform distributions. Their PDFs are:

$$f_{S_{sy}}\left(s_{sy}\right) = \begin{cases} \dfrac{1}{18.57} & 21.21 \leq s_{sy} \leq 39.67 \\ 0 & \text{otherwise.} \end{cases}$$

$$f_{\tau_D}\left(\tau_D\right) = \begin{cases} \dfrac{1}{7.73} & 15.67 \leq s_{sy} \leq 23.4 \\ 0 & \text{otherwise.} \end{cases}$$

Use the interference theory to calculate the reliability of the pin.

3.5.   The failure mode of a bar under axial loading is a static failure at which the maximum normal stress σ_{max} at the critical section is more than the ultimate material strength S_u. The ultimate material strength S_u follows a normal distribution with a mean $\mu_{S_u} = 78.2$ (ksi) and standard deviation $\sigma_{S_u} = 6.9$ (ksi). The maximum normal stress σ_{max} at the critical section is a normally distributed random variable with a mean $\mu_{\sigma_{max}} = 57.4$ (ksi) and standard deviation $\sigma_{\sigma_{max}} = 10.21$ (ksi). Use the interference theory to calculate the reliability of the bar.

3.6. The maximum bending stress σ_{max} (ksi) at the critical section of a beam follows a log-normal distribution with a mean $\mu_{\ln \sigma_{max}} = 3.22$ and standard deviation $\sigma_{\ln \sigma_{max}} = 0.52$. The material yield strength S_y (ksi) follows a log-normal distribution with a mean $\mu_{\ln S_y} = 4.56$ and standard deviation $\sigma_{\ln S_y} = 0.17$. It is assumed that the component is a failure when the maximum bending stress σ_{max} is more than the material yield strength S_y. Use the interference theory to calculate the reliability of the component.

3.7. A bar is under three statistically independent axial loadings P_1, P_2, and P_3. These three axial loadings follow normal distributions. Their means and standard deviations are:

$\mu_{P_1} = 1.51$ (kip), $\sigma_{P_1} = 0.121$ (kip; $\mu_{P_2} = 3.79$ (kip), $\sigma_{P_2} = 0.292$ (kip) and $\mu_{P_3} = 14.12$ (kip), $\sigma_{P_3} = 1.32$ (kip). The maximum normal stress at the critical section is the sum of the normal stress induced by these three axial-loadings, that is, $\sigma_A = 2.264 P_1 + 2.264 P_2 + 2.264 P_3$ (ksi). The yield strength S_y of the material is a normally distributed random variable with the mean $\mu_{S_y} = 61.5$ (ksi) and the standard deviation $\sigma_{S_y} = 5.91$ (ksi). Use the FOSM method to calculate the reliability of the bar.

3.8. The bar is under axial loading. The diameter d of the bar can be treated as a normal distribution with a mean $\mu_d = 0.75$ (in) and a standard deviation $\sigma_d = 0.002$ (in). The axial loading P is a normal distribution with a mean $\mu_P = 8.12$ (kip) and a standard deviation $\sigma_P = 2.45$ (kip). The yield strength S_y of the material is a normally distributed random variable with a mean $\mu_{S_y} = 61.5$ (ksi) and a standard deviation $\sigma_{S_y} = 5.91$ (ksi). Use the FOSM method to calculate the reliability of the bar.

3.9. A shaft is under a torsion. The diameter d of the shaft can be treated as a normal distribution with a mean $\mu_d = 1.125$ (in) and a standard deviation $\sigma_d = 0.002$ (in). The torsion T is a normal distribution with the mean $\mu_T = 2.89$ (kip) and the standard deviation $\sigma_T = 0.24$ (kip). The shear yield strength S_{sy} of the material is a normal distributed random variable with a mean $\mu_{S_{sy}} = 31.5$ (ksi) and a standard deviation $\sigma_{S_{sy}} = 2.98$ (ksi). Use the FOSM method to calculate the reliability of the shaft.

3.10. A beam with a constant round cross-section is under a bending moment. The diameter d of the shaft can be treated as a normal distribution with a mean $\mu_d = 1.25$ (in) and a standard deviation $\sigma_d = 0.002$ (in). The bending moment M on the critical section is a normal distribution with a mean $\mu_M = 2.25$ (kip) and a standard deviation $\sigma_M = 0.19$ (kip). The yield strength S_y of the material is a normal distributed random variable with a mean $\mu_{S_y} = 61.5$ (ksi) and a standard deviation $\sigma_{S_y} = 5.91$ (ksi). Use the FOSM method to calculate the reliability of the beam.

3.11. Use the H-L method to compile the MATLAB program for calculating the reliability of Problem 3.9.

3.12. Use the H-L method to compile the MATLAB program for calculating the reliability of Problem 3.10.

3.13. A component is under axial tensile loading with a limit state function $g\left(S_y, \sigma_N\right) = S_y - \sigma_N$. S_y is the material yield strength, which follows a normal distribution with the mean $\mu_{S_y} = 61.5$ (ksi) and the standard deviation $\sigma_{S_y} = 5.91$ (ksi). The normal stress σ_N (ksi) of the component at the critical section is a log-normal distributed variable with the log-mean $\mu_{\ln \sigma_N} = 3.9$ and the log-standard deviation $\sigma_{\ln \sigma_N} = 0.159$. Use the R-F method to compile the MATLAB program for calculating the reliability of this component.

3.14. A component is under axial tensile loading. The component strength index is the ultimate tensile strength S_u. The ultimate tensile strength S_u (ksi) follows a Weibull distribution with $\eta = 47$ and $\beta = 1.5$. The axial loading P at the critical section is a normal distributed random variable with a mean 18.6 (klb) and a standard deviation 2.1 (klb). The diameter d of the critical section follows a normally distributed variable with a mean 1.750″ and a standard deviation 0.002″. Build the limit state function and then use the R-F method to determine the reliability of the component.

3.15. Calculate the trial number N with a 95% confidence level for a critical component with a pre-specified relative error 1% of the probability of component failure.

3.16. If the component is treated as a key component, use the Monte Carlo method to calculate the reliability of Problem 3.13. Estimate the range of reliability with a 95% confidence level.

3.17. If the component is treated as a key component, use the Monte Carlo method to calculate the reliability of Problem 3.14. Estimate the range of the reliability with a 95% confidence level.

3.18. Use Monte Carlo method to calculate the reliability of Problem 3.9.

3.19. Use Monte Carlo method to calculate the reliability of Problem 3.10.

C H A P T E R   4

Reliability of a Component under Static Load

4.1   INTRODUCTION

According to the applied loading on a component, component design typically can be grouped into component design under static loading and component design under cyclic loading. This chapter will discuss component design under static loading. The component design under cyclic loading will be discussed in the book of *Reliability-Based Mechanical Design, Volume 2: Component under Cyclic Load and Dimension Design with Required Reliability.*

According to the task of component design, component design can be grouped into component design check and component dimension design. Component design check means that we check the actual reliability of a component when loading, geometric dimensions, and material properties are all specified. If the actual reliability of the component is larger than the specified reliability, the component is classified as a qualified design. Otherwise, the component needs to be redesigned. Component dimension design refers that we design component dimensions with the required reliability under the specified loading and material. This chapter will discuss component design check for a component under static loading. The component dimension design will be discussed in the book of *Reliability-based Mechanical Design, Volume 2: Component under Cyclic Load and Dimension Design with Required Reliability.*

In this chapter, first, we will discuss how to describe all design parameters: dimension, loading, and material properties as random variables. Then we will discuss the reliability of component under some typical static loadings such as rod under axial loading, pin under direct shearing, shaft under torsion, beam under bending. Finally, we will discuss component under combined loadings.

4.2   GEOMETRIC DIMENSION AS A RANDOM VARIABLE

Any geometric dimension is always associated with a corresponding dimension tolerance, which is controlled by a manufacturing process. For example, the possible dimension tolerance of a component dimension between $0.4''-1.97''$ vs. its different manufacturing processes [1] are listed in Table 4.1.

The geometric dimensions of a component can be classified as a free dimension and a mating dimension. A free dimension refers that the variation of the dimension will not affect

Table 4.1: Dimension tolerance vs. its different manufacturing processes

Dimension Range	0.4″–0.71″	0.71″–1.19″	1.19″–1.97″
Manufacturing Process	Tolerances in Thousandths of an Inch		
Lapping and honing	0.2–0.3	0.25–0.4	0.3–0.4
Cylindrical grinding	0.3–0.7	0.4–0.8	0.4–1.0
Surface grinding	0.3–1.0	0.4–1.2	0.4–1.6
Diamond turing	0.3–0.7	0.4–0.8	0.4–1.0
Diamond boring	0.3–0.7	0.4–0.8	0.4–1.0
Broaching	0.3–1.0	0.4–1.2	0.4–1.6
Reaming	0.4–2.8	0.5–3.5	0.6–4.0
Turing	0.7–10	0.8–10	1.0–10
Boring	1.0–10	1.2–10	1.6–10
Milling	2.8–10	2.5–12	4.0–16
Planing and shaping	2.8–10	2.5–12	4.0–16
Drilling	2.8–10	2.5–12	4.0–16

component functions and other adjacent components. The dimension tolerance of a free dimension will typically be the default tolerance, which is the most economical way to manufacture this dimension in a company. A schematic of a simple base-shaft assembly is shown in Figure 4.1, where a base is a rectangular place with a hole in the center and the shaft is a cylindrical round bar. The shaft is fitted through the hole of the base. The dimension $d1$ in Figure 4.1 is a free dimension.

A mating-dimension refers that the variation of this dimension will affect component functions and other adjacent components. The mating dimension is the dimension on an interface. The tolerance of it will be determined by the required function, such as a clearance fit, interference fit, or transient fit. The diameter $d2$ of the shaft and the diameter $d2$ of the hole will be a mating dimension.

The dimension tolerance can be expressed with symmetrical tolerance such as $1.000″ \pm 0.005$, or a unilateral tolerance such as or a bilateral tolerance such as $1.000″^{+0.008}_{-0.002}$. The general form of a dimension with tolerance can be expressed as $d^{t_U}_{t_L}$, where d is the nominal dimension, t_U is the upper limit of tolerance and t_L is the lower limit of tolerance. If the final value of the dimension of a manufactured component is in the range $(d + t_L, d + t_U)$, the component will be classified as a qualified component. If the final value of the dimension of a manufactured component is out of the range $(d + t_L, d + t_U)$, the component will be classified as an unqualified component and will be discarded.

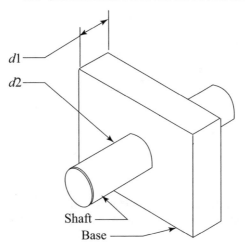

Figure 4.1: The schematic of a base-shaft assembly.

Before a component has been manufactured, the actual dimension of a component is un-known due to the associated dimension tolerance. Therefore, component geometric dimension is a random variable. It is typically treated as a normally distributed random variable d. According to the definition of dimension tolerance, the components' dimension inside the dimension tolerance range $(d + t_L, d + t_U)$ will be accepted. For a normal distribution, the probability of event $(\mu_d - 4\sigma_d \leq d \leq \mu_d + 4\sigma_d)$ will be 99.9968%. This event can be used to represent the dimension tolerance range with a very small error (0.0032%). Therefore, the mean and standard deviation of a normally distributed dimension random variable d can be determined per Equation (4.1):

$$\mu_d = \frac{(d + t_L) + (d + t_U)}{2} = d + \frac{t_L + t_U}{2}$$
$$\sigma_d = \frac{(d + t_U) - (d + t_L)}{8} = \frac{t_U - t_L}{8}. \tag{4.1}$$

Example 4.1
A shaft diameter with a dimension tolerance is $1.250'' \pm 0.005$. Determine the mean and standard deviation of the shaft diameter if it is treated as a normally distributed random variable.

Solution:
Per Equation (4.1), the mean and standard deviation of the shaft diameter are

$$\mu_d = d + \frac{t_L + t_U}{2} = 1.250 + \frac{-0.005 + 0.005}{2} = 1.250''$$

$$\sigma_d = \frac{t_U - t_L}{8} = \frac{0.005 - (-0.005)}{8} = 0.00125''.$$

■

Example 4.2

The diameter of a hole with a dimension tolerance is $1.500''^{0.009}_{0.001}$. Determine its mean and standard deviation of the hole diameter if it is treated as a normally distributed random variable.

Solution:

Per Equation (4.1), the mean and the standard deviation of the shaft diameter are

$$\mu_D = d + \frac{t_L + t_U}{2} = 1.500 + \frac{0.001 + 0.009}{2} = 1.505''$$

$$\sigma_D = \frac{t_U - t_L}{8} = \frac{0.009 - (-0.001)}{8} = 0.001''.$$

■

4.3   STATIC LOADING AS A RANDOM VARIABLE

In traditional mechanical component design under static loading, the maximum static loading will be one unique value. However, in reliability-based mechanical component design, the loading for component design under static loading is a random variable. In reliability-based mechanical design, a component under consideration is not a single specific component, but a batch of the same component. Each of this batch of the component might be used by different customers in different service conditions. One value of the maximum loading for each component can be obtained. This value can be treated as one sample data of the static loading. Suppose that this batch of the component under consideration is 50,000. We could collect one sample data per each component. Therefore, we will have 50,000 sample data of the static loading for this batch of the same component. Based on these sample data, we can determine the type of distribution and calculate the corresponding distribution parameters of the static loading.

If a loading P such as concentrated force, concentrated moment or torque, is expressed as a range of value such as (P_{low}, P_{up}), it could be treated as a normally distributed random variable. We can use the same reasoning and similar equation as Equation (4.1) to determine its mean and standard deviation:

$$\mu_P = \frac{\left(P_{low} + P_{up}\right)}{2}$$

$$\sigma_P = \frac{\left(P_{up} - P_{low}\right)}{8}.$$

(4.2)

For a component design under static loading, loading could also be presented by other three types: (1) one single loading, such as 2000 lb; (2) a discrete distributed random variable such as the loading in Example 4.4; and (3) a continuously distributed random variable such as the loading in Example 4.5.

Example 4.3
A component is subjected to a concentrated force $P = 2025 \pm 125$ (lb). If it is treated as a normal distribution, determine its mean and standard deviation.

Solution:
Per Equation (4.2), the mean and standard deviation of this concentrated force are:

$$\mu_P = \frac{\left(P_{low} + P_{up}\right)}{2} = \frac{[(2025 - 125) + (2025 + 125)]}{2} = 2025 \,(\text{lb})$$

$$\sigma_P = \frac{\left(P_{up} - P_{low}\right)}{8} = \frac{[(2025 + 125) - (2025 - 125)]}{8} = 31.25 \,(\text{lb}).$$

∎

Example 4.4
A company will design a component. The company sent out a survey to potential customers for identifying the maximum static loading for the component. The collected data of maximum static loading for the batch of the same component are listed in Table 4.2. In this table, the first row is the maximum possible loading for the component. The second row shows the number of selected loading by potential customers. For example, in the fourth column, "1100" in the first row refers that the maximum static loading is 1100 lb. The "40" in the second row refers that 40 customers in the survey selected 1100 lb as the maximum static loading. Build the distribution function for the loading.

Table 4.2: The collected data for the maximum static loading

Maximum static loading L (lb.)	1000	1050	1100	1150	1200
Number of selected loading	10	20	40	32	5

Solution:

In this example, the maximum static loading, according to the survey, is a discrete random variable. The total number of sample data will be $10 + 20 + 40 + 32 + 5 = 107$. Then the probability of the loading equal to 1100 will be 40/107. So, we can use the following PMF to describe the loading:

$$p\,(l) = \begin{cases} 10/107 = 0.0935 & l = 1000\,(\text{lb}) \\ 20/107 = 0.1869 & l = 1050\,(\text{lb}) \\ 40/107 = 0.3738 & l = 1100\,(\text{lb}) \\ 32/107 = 0.2991 & l = 1150\,(\text{lb}) \\ 5/107 = 0.0467 & l = 1200\,(\text{lb}). \end{cases}$$

∎

Example 4.5

A company collected maximum static loading for a component. One hundred components at different services were selected for collecting loading data. For each component, the maximum loading on the component was collected during one-month regular service. The collected data are listed in Table 4.3. Create the histogram and then determine the type of distribution and corresponding distribution parameters.

Table 4.3: One hundred sample data of the maximum static loading on the component

The Maximum Static Loading X on Components (lb)
2532, 1987, 2877, 2588, 2511, 2682, 2568, 2026, 2377, 2384, 2211, 3207, 2056, 2232, 2130, 1922, 2363, 2327,
3139, 2338, 1971, 3172, 3006, 2347, 1772, 2250, 2380, 2574, 2332, 2650, 2626, 1887, 2023, 2117, 2221, 2306,
2456, 1087, 2244 ,3009, 1970, 2870, 2608, 2437, 2532, 1746, 2412, 3172, 2494, 2469, 2120, 2436, 2555, 2642,
2282, 2344, 3361, 1434, 3453, 2602, 2900, 1701, 2184, 2325, 2640, 1698, 2662, 1904, 2480, 2744, 2597, 2937,
2903, 2157, 2566, 2025, 1855, 2866, 2450, 2425, 2860, 2718, 2608, 3013, 2868, 2558, 2139, 2157, 2986, 1725,
2439, 1573, 2909, 2838, 2451, 2418, 1331, 2712, 1463, 1406,

Solution:

1. Use MATLAB to create a histogram.

   We can follow Section 2.8 and use the MATLAB program to create the histogram. We need to input the data of from the Table 4.3 in an Excel file, where the data will be inputted in the first column. The file name will be "Example 4.5." Then we can use the following MATLAB program to import the data and create the histogram. The histogram is displayed in Figure 4.2. From the histogram, we could assume that the loading follows a normal distribution.

```
%Import the data from an Excel file
X=xlsread(`Example 4.5}'),
% Create a 10-bin histogram
histogram(X,10)
xlabel (`Bins')
ylabel(`Frequency')
```

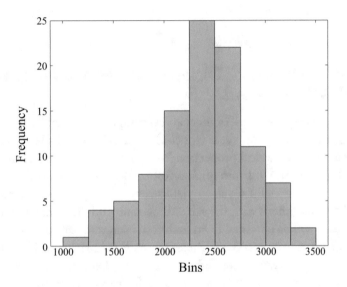

Figure 4.2: Histogram of the maximum static loading.

2. Run the goodness-of-fit test and calculate the distribution parameters.

We can follow Section 2.13 to check whether it is a normal distribution. We can use the following MATLAB program to run the goodness-of-fit test. After running the MAT-LAB program, we have the value of "h" is zero. So the load can be assumed as a normal distribution. The MATLAB program can also calculate the mean and standard deviation:

$$\mu_X = 2400.2 \,(\text{lb}), \qquad \sigma_X = 457.4 \,(\text{lb}).$$

```
% The goodness-of-fit test on the data in the excel
% file "Example 4.5.xls".
X=xlsread('Example4.4');
Q2=fitdist(X,'norm')
h=chi2gof(X,'CDF',Q2)
```

■

4.4   MECHANICAL PROPERTIES OF MATERIALS AS RANDOM VARIABLES

Most material mechanical properties for mechanical component design typically come from two types of testing: tensile testing and shear (torsion) testing. From tensile testing, we can obtain four important material mechanical properties: Young's modulus E, yield strength S_y, ultimate strength S_u, and Poisson ratio ν. From torsion testing, we can obtain three important material properties: shear Young's modulus G, shear yield strength S_{sy}, and ultimate shear strength S_{su}.

For a set of test specimens with the same dimension and same material, different material mechanical properties will be obtained from each test; even tests are conducted on the same test equipment with the same test procedure. Therefore, material mechanical properties are random variables because there are always some slight variations in chemical composition and some variations in heat treatment and manufacturing process for the same brand name of the material. Another important cause for this is that there are always some randomly distributed "defects" inside materials such as voids and dislocation [2]. After test data of material mechanical properties are collected, we can determine their types of distributions and corresponding distribution parameters. Typically, we can use a normal distribution or a log-normal distribution, or a Weibull distribution to describe material mechanical properties.

Table 4.4 displays mean, standard deviation, and coefficient of variance of Young's modulus E, shear Young's modulus G, and Poisson ratio ν of some materials from the literature [3]. In this table, Young's modulus E, shear Young's modulus G, and Poisson ratio ν all follow a normal distribution. In the second row of this table, μ_X, σ_X, and γ_X are the mean, standard deviation, and coefficient of variance of a normally distributed random variable X, respectively. The subscript X in the second row can be Young's modulus E, the shear Young's modulus G, and the Poisson ratio ν.

Dr. E. B. Haugen published a book [4] in 1980 and provided distribution parameters of yield strength and ultimate strengths of some materials. Following Tables 4.5–4.10 are a small selected data form this book. In these tables, μ_{S_u}, σ_{S_u}, and γ_{S_u} are the mean, standard deviation, and coefficient of variance of ultimate strength, respectively. μ_{S_y}, σ_{S_y}, and γ_{S_y} are the mean, standard deviation, and coefficient of variance of yield strength, respectively. Those

Table 4.4: Young's modulus, shear modulus, and Poisson's ratio of some materials

Mechanical Properties	E (MPa)			G (Mpa)			v (Poisson's Ratio)		
Materials	μ_E	σ_E	γ_E	μ_G	σ_G	γ_G	μ_v	σ_v	γ_v
Steel	206010	3269.7	0.0159	78970.5	163.5	0.0021	0.29	0.01333	0.046
Alloy steel	201105	4905	0.0244	79461	/	/	0.285	0.015	0.0526
Grey iron	134888	7357.5	0.0545	44145	/	/	0.25	0.00667	0.0267
Ductile iron	142245	4905	0.0345	73084.5	538.7	0.006	/	/	/
Aluminum and its alloy	69651	3269.7	0.0469	25996.5	163.5	0.0063	0.3333	/	/
Copper and its alloy	100062	9165	0.0916	42183	98.1	0.0023	0.365	0.01838	0.0502
Titanium alloy	112315	1635.3	0.0145	40858.7	1336.7	0.0327	0.30667	0.01155	0.0377

material properties all follow a normal distribution. Table 4.5 lists distribution parameters of ultimate strength and yield strength of some aluminum alloy. Table 4.6 displays distribution parameters of ultimate strength and yield strength of some Magnesium alloys and titanium alloys. Table 4.7 displays distribution parameters of ultimate strength and yield strength of some steels. Table 4.8 displays distribution parameters of ultimate strength and yield strength of some steel alloys. Table 4.9 displays distribution parameters of ultimate strength and yield strength of some stainless steels. Table 4.10 displays distribution parameters of ultimate strength and yield strength of some irons.

4.5 ESTIMATION OF SOME DESIGN PARAMETERS

In reliability-based mechanical design, reliability is the measure of a component safety status. The reliability of a component is solely determined by and based on the statistical descriptions of all design parameters, that is, the type of distributions and corresponding distribution parameters of all design parameters. We should use the reliable distribution parameters based on reliable statistical data to conduct the calculation of the reliability. However, for a rough estimation of the reliability, when the statistical descriptions of material mechanical properties and some design

Table 4.5: Ultimate strength and yield strength of some aluminum alloy

Materials	Remarks	Ultimate Strength S_u (ksi)				Yield Strength S_y (ksi)			
		μ_{S_u}	σ_{S_u}	γ_{S_u}	Sample size	μ_{S_y}	σ_{S_y}	γ_{S_y}	Sample size
2014	Plate, L-T, 0.50–1.5″	85.0	2.76	0.032	20	75.0	1.75	0.023	20
2014-T651	Plate L-T, Ambient temperature	69.0	1.82	0.026	20	63.0	1.35	0.021	20
2024-T81	Clad sheet, 0.068–0.140 L-T	69.9	1.70	0.024	15	64.4	2.41	0.037	15
2219-T87	Plate, 0.50–1.5″ L-T	67.0	1.64	0.024	17	54.0	1.53	0.028	17
6061-T4	Sheet, 0.032–0.125″	36.6	1.69	0.046	1461	20.1	2.30	0.114	1461
6061-T6	Sheet, 0.032–0.125″	45.6	1.91	0.042	1648	41.7	2.89	0.069	1648
2219-T87	Plate, 0.50–1.5″ T-L	67.0	1.64	0.024	17	54.0	1.53	0.028	17
7075-T6	Clad sheet, 0.125 L-T	76.2	1.08	0.014	300	65.3	1.52	0.023	300

Table 4.6: Ultimate strength and yield strength of some magnesium alloys and titanium alloys

Materials	Remarks	Ultimate Strength S_u (ksi)				Yield Strength S_y (ksi)			
		μ_{S_u}	σ_{S_u}	γ_{S_u}	Sample size	μ_{S_y}	σ_{S_y}	γ_{S_y}	Sample size
Magnesium Alloys									
AZ 31-B-H24	Sheet and plate 0.016–0.249 L-T	41.2	0.95	0.023	/	31.2	0.95	0.030	/
AZ 91-T4	Castings QQ-M-56	38.5	2.26	0.059	30	14.2	0.67	0.047	30
AZ 318-H24	Sheet and plate 0.0016–0.249	41.78	0.97	0.023	14800	31.54	1.34	0.042	14800
Titanium Alloys									
Titanium 99%	Commercially pure, L-T	101.9	6.98	0.068	1363	85.1	9.03	0.106	1354
T1-6Al-4V	Sheet and bar Annealed 75°F	135.5	6.7	0.049	2542	130.6	7.2	0.055	2611
T1-6Al-4V	Sheet and bar Temperature 572°F	99.1	4.49	0.045	462	81.0	5.25	0.065	556
T1-6Al-4V	Sheet and bar Forgings	143	4.98	0.035	89	135.3	4.75	0.035	89
T1-6Al-4V	Sheet, heat treated and aged	175.4	7.91	0.045	603	163.7	9.03	0.055	603

Table 4.7: Ultimate strength and yield strength of some steels

Materials	Remarks	Ultimate Strength S_u (ksi)				Yield Strength S_y (ksi)			
		μ_{S_u}	σ_{S_u}	γ_{S_u}	Sample size	μ_{S_y}	σ_{S_y}	γ_{S_y}	Sample size
C1006	Sheet, hot rolled	48.3	0.52	0.011	5	35.7	0.8	0.022	5
C1018	Round bar, cold draw	87.6	5.74	0.066	50	78.4	5.9	0.075	50
C1035	Round bar, hot rolled, 1-9″ diameter	86.2	3.92	0.045	913	49.5	5.36	0.108	899
C1045	Round bar, cold drawn 3/4–1 1/4″ diameter	117.7	7.13	0.061	30	95.5	6.59	0.069	25
Low-carbon steel	Sheet, drawing quality, hot rolled, 0.075″	44.7	1.26	0.028	140	34	2.25	0.066	140
Low-carbon steel	Sheet, drawing quality, cold rolled	44.7	1.12	0.025	140	25.6	1.87	0.073	140
Low-carbon steel	Casting, as cast	66.7	2.01	0.030	200	34.8	0.84	0.024	200

Table 4.8: Ultimate strength and yield strength of some steel alloys

Materials	Remarks	Ultimate Strength S_u (ksi)				Yield Strength S_y (ksi)			
		μ_{S_u}	σ_{S_u}	γ_{S_u}	Sample size	μ_{S_y}	σ_{S_y}	γ_{S_y}	Sample size
5Cr-Mo-V alloy	5Cr-1.5Mo-0.4V-0.35C 3/4″ dia bar aircraft steel	296.5	1.84	0.006	25	240.8	3.1	0.013	25
Nickel steel	0.2C-2.25Ni casting, normalized and tempered	83.2	2.35	0.028	200	55.4	1.54	0.028	200
Ni-Cr-Mo alloy	0.28C-2Mi-0.8Cr-0.35Mo, normalized, quenched and tempered	161.1	4.97	0.031	44	148.1	4.98	0.034	44
High-strength structural steel	0.22C-1.35Mn-0.025O 0.50–1.50 in	76.9	2.06	0.027	109	49.6	3.69	0.074	113
High-strength structural steel	0.25Si-0.25Cu as rolled, 0.134-4″	79.1	4.65	0.059	214	51	4.63	0.091	234
ASTM A7	Plate (heavy) ambient temp.	64.8	2.93	0.045	141	39.8	4.15	0.104	141
ASTM A7	Structural steel angles, channels, beams	65.6	2.43	0.037	166	42.7	4.86	0.114	166

Table 4.9: Ultimate strength and yield strength of some stainless steels

Materials	Remarks	Ultimate Strength S_u (ksi)				Yield Strength S_y (ksi)			
		μ_{S_u}	σ_{S_u}	γ_{S_u}	Sample size	μ_{S_y}	σ_{S_y}	γ_{S_y}	Sample size
Type 202	Strip, austenitic cool rolled 0.017–0.022"	99.7	2.71	0.027	25	49.9	1.32	0.026	25
Type 301	Strip and sheet annealed, 0.038–0.195"	105	5.68	0.054	100	46.8	4.7	0.100	100
Type 304	Round bars annealed, 0.50-4.625"	85	4.14	0.049	45	37.9	3.76	0.099	35
Type 403	Bar, ferritic hot rolled	109.7	4.48	0.041	549	81.2	6.86	0.084	549
Type 347	Forging -100F	118	3.34	0.028	23	34.5	1.15	0.033	187
AM-350	Sheet annealed, 0.025–0.125"	152.8	6.7	0.044	194	63.4	4.04	0.064	188

parameters are not available due to lack of sample data, we can use the following methods to estimate their distribution parameters and then estimate the reliability of a component.

1. Estimate ultimate shear strength by using ultimate tensile strength.

   Material mechanical properties from tensile tests are typically available from design handbooks. We can use the following Equation (4.3) [4] to estimate ultimate shearing strength by using ultimate tensile strength:

$$\frac{S_{su}}{S_u} = \begin{cases} 0.60 & \text{Aluminum alloy} \\ 0.75 & \text{Steel} \\ 0.90 & \text{Copper} \\ 0.90 & \text{Malleable iron} \\ 1.30 & \text{Cast iron,} \end{cases} \qquad (4.3)$$

where S_u is ultimate tensile strength and S_{su} is ultimate shear strength.

2. Estimate shear yield strength by using tensile yield strength.

Table 4.10: Ultimate strength and yield strength of some irons

Materials	Remarks	Ultimate Strength S_u (ksi)				Yield Strength S_y (ksi)			
		μ_{S_u}	σ_{S_u}	γ_{S_u}	Sample size	μ_{S_y}	σ_{S_y}	γ_{S_y}	Sample size
Malleable	Ferrritic foundary A	53.4	2.68	0.050	434	34.9	1.47	0.042	434
Iron grade 32510	Grade 32510 foundary B	56	1.41	0.025	785	39.1	1.22	0.031	785
Malleable iron	Pearilitic 217-233 BHN spherodized	93.9	4.33	0.046	172	61.8	2.48	0.040	172
Nodular iron grade 60-45-10	As cast grade 60-45-10	73.2	5.37	0.073	412	54.5	4.65	0.085	412
Nodular iron	As cast auto crankshafts	99.3	6.14	0.062	125	74.9	4.62	0.062	125

The following Equation (4.4) [4] can be used to estimate shearing yield strength by using tensile yield strength:

$$\frac{S_{sy}}{S_y} = \begin{cases} 0.5 & \text{based on the Maximum shear stress failure theory} \\ 0.577 & \text{based on distortion—energy failure theory,} \end{cases} \quad (4.4)$$

where S_y is tensile yield strength and S_{sy} is shear yield strength.

3. Mean and standard deviation if the range of material mechanical properties are given.

If the range of a material mechanical property X is given, we can assume that X follows a normal distribution with a mean μ_X and a standard deviation σ_X. We can use the following equations to estimate the mean and the standard deviation of the material mechanical property X:

$$\mu_X = \frac{X_{\min} + X_{\max}}{2}$$
$$\sigma_X = \frac{X_{\max} - X_{\min}}{8}, \quad (4.5)$$

where $X_{\min}$ and $X_{\max}$ are the lower and upper limits of the material mechanical property X, respectively.

4. Mean and standard deviation if only the average of material mechanical property is given.

For a material mechanical strength such as yield strength or ultimate strength, if only an average value is available, we assume that it follows a normal distribution and could use the following approach to estimate its mean and standard deviation:

$$\mu_X = X_{average}$$
$$\gamma_X = 0.1 \qquad\qquad (4.6)$$
$$\sigma_X = \gamma_X \times \mu_X = 0.1 X_{average},$$

where $X_{average}$ is the average of the sample data of the material mechanical property. γ_X is the coefficient of variance. μ_X and σ_X is the mean and the standard deviation of its normal distribution.

For Young's modulus E, shear Young's modulus G and the Poisson ratio v, if their averages are known, we can assume that they follow a normal distribution and use following equations [6] to estimate their mean and standard deviation.

For Young's modulus E,

$$\gamma_E = 0.01$$
$$\mu_E = E_{average} \qquad\qquad (4.7)$$
$$\sigma_E = \gamma_E \times \mu_E = 0.01 E_{average},$$

where γ_E is the coefficient of variance of the material Young's modulus. $E_{average}$ is the average of sample data of the material Young's modulus. μ_E and σ_E are the mean and the standard deviation of normally distributed material Young's modulus.

For shear Young's modulus G,

$$\gamma_G = 0.025$$
$$\mu_G = G_{average} \qquad\qquad (4.8)$$
$$\sigma_G = \gamma_G \times \mu_G = 0.025 G_{average},$$

where, γ_G is the coefficient of variance of the material shear Young's modulus. $G_{average}$ is the average of sample data of the material shear Young's modulus. μ_G and σ_G are the mean and the standard deviation of normally distributed material shear Young's modulus.

For Poisson ratio v,

$$\gamma_v = 0.042$$
$$\mu_v = v_{average} \qquad\qquad (4.9)$$
$$\sigma_v = \gamma_v \times \mu_v = 0.042 v_{average},$$

where γ_v is the coefficient of variance of the material Poisson ratio. $v_{average}$ is the average of sample data of the material Poisson ratio. μ_v and σ_v are the mean and the standard deviation of normally distributed material Poisson ratio.

5. The stress concentration factor.

The stress concentration factor is a function of geometric shape and dimension. Since the geometric dimension is a random variable, the stress concentration factor is also a random variable and typically follows a normal distribution. We can use the following equations to determine the mean and the standard deviation of stress concentration factor [6, 7]:

$$\gamma_K = 0.05$$
$$\mu_K = K_{Table} \tag{4.10}$$
$$\sigma_K = \gamma_K \times \mu_K = 0.05 K_{Table},$$

where γ_K is the coefficient of variance of the stress concentration factor. K_{Table} is the stress concentration factor obtained from tables in current design handbooks or design books. μ_K and σ_K are the mean and the standard deviation of normally distributed stress concentration factor.

Example 4.6
A material with only three tensile tests has the following mechanical properties, as shown in Table 4.11. If all mechanical properties are assumed to be normal distributions, estimate their distribution parameters of this material mechanical properties.

Table 4.11: Average values from three tensile tests

$S_{y\text{-}average}$ (ksi)	$S_{u\text{-}average}$ (ksi)	$E_{average}$ (ksi)	$v_{average}$
49.2	61.3	2.91×10^4	0.282

Solution:
We can use Equation (4.6) to estimate the mean and standard deviation of the yield strength and the ultimate strength. Per Equations (4.7) and (4.9), we can estimate the mean and the standard deviation of Young's modulus and Poisson ratio, respectively.

We can use the relationship ration $G = \dfrac{E}{2(1 + v)}$ among Young's modulus, shear Young's modulus, and Poisson ratio to calculate the shear Youngs modulus, and then use Equation (4.8) to estimate the mean and standard deviation of the shear Young's modulus.

We can use Equations (4.3) and (4.4) to estimate the ultimate shear strength and the shear yield strength, respectively, and then use Equation (4.6) to estimate their mean and standard deviation.

All of these estimations are listed in Table 4.12.   ∎

Table 4.12: The estimated mechanical properties for Example 4.6

Yield strength S_y (ksi)		Ultimate strength S_u (ksi)		Youngs' Modulus E (ksi)		Poisson ratio: v	
μ_{S_y}	σ_{S_y}	μ_{S_u}	σ_{S_u}	μ_E	σ_E	μ_v	σ_v
49.2	4.92	61.3	6.13	2.91×10^4	291	0.282	0.0118
Shear yield strength S_{sy} (ksi)		Shear ultimate strength S_{su} (ksi)		Shear Young's modulus G (ksi)			
$\mu_{S_{sy}}$	$\sigma_{S_{sy}}$	$\mu_{S_{su}}$	$\sigma_{S_{su}}$	μ_G	σ_G		
24.6	2.46	45.98	4.60	1.13×10^4	284		

4.6   RELIABILITY OF A ROD UNDER AXIAL LOADING

4.6.1   RELIABILITY OF A ROD UNDER AXIAL LOADING FOR A STRENGTH ISSUE

For calculating the reliability of a component under axial loadings, we need to establish the limit state function of the component under such loading.

A rod under axial loading will produce normal axial stress. The following equation can calculate the normal stress of the rod:

$$\sigma = \frac{F_a}{A},\qquad(4.11)$$

where σ is the normal stress on the cross-section induced by the axial loading. F_a and A are the resultant axial force on and the cross-section area of the component critical section.

For a component made of brittle material, the component under axial loading will be treated as a failure when the maximum normal stress is larger than the ultimate tensile strength. The limit state function of this case will be:

$$g\left(S_u, K_t, A, F_a\right) = S_u - K_t\frac{F_a}{A} = \begin{cases} > 0 & \text{Safe} \\ = 0 & \text{Limit state} \\ < 0 & \text{Failure,} \end{cases}\qquad(4.12)$$

where S_u is the ultimate tensile strength of component material. K_t is the stress concentration factor of the component critical section. The meanings of F_a and A are the same as those in Equation (4.11).

For a component made of ductile material, the component under axial loading will be treated as a failure when the maximum normal stress is larger than the yield strength. The limit state function of this case will be:

$$g\left(S_y, K_t, A, F_a\right) = S_y - K_t\frac{F_a}{A} = \begin{cases} > 0 & \text{Safe} \\ = 0 & \text{Limit state} \\ < 0 & \text{Failure,} \end{cases} \qquad (4.13)$$

where S_y is the yield strength of component material. F_a, A, and K_t have the same meanings as those in Equation (4.12).

When a component's axial loading can be described by a PDF, that is, a continuous random variable, we can use Equations (4.12) and (4.13) to calculate the reliability of component by using the techniques discussed in Chapter 3.

When a component's axial loading is described by a PMF, we can use the total probability theorem Equation (2.24) in Chapter 2. In this case, the axial loading is expressed by a PMF as:

$$P\left(F_a = F_i\right) = p_i \qquad i = 1, \ldots, n, \qquad (4.14)$$

where p_i is a probability of loading F_a when it is equal to F_i. Since it is a PMF, we have:

$$\sum_{i=1}^{n} p_i = 1. \qquad (4.15)$$

The reliability of the component under such axial loading will be:

$$R = \sum_{i=1}^{n} \left(p_i \times R_i\right), \qquad (4.16)$$

where R is the reliability of component under such axial loadings. R_i is the reliability of component when the axial loading F_a is equal to F_i. Even F_i can be a constant value, that is, a deterministic value, we still need to use Equations (4.12) or (4.13) to calculate R_i since the cross-section area A and the stress concentration factor K_t are random variables.

Three examples will demonstrate how to calculate the reliability of components under axial loading.

Example 4.7
A beam ABC with a link bar CD is subjected to a concentrated force $P = 1.90 \pm 0.180$ (klb) at point B, as shown in Figure 4.3. The link bar CD is a solid round bar with a diameter $d = 0.250 \pm 0.005''$. The bar is made of a ductile material. The yield strength S_y of the bar's material follows a normal distribution with a mean $\mu_{S_y} = 34.5$ (ksi) and a standard deviation $\sigma_{S_y} = 3.12$ (ksi). Calculate the reliability of the link bar CD.

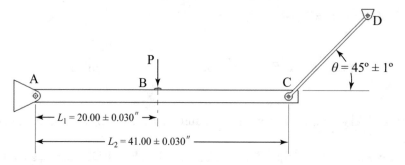

Figure 4.3: A beam *ABC* with a link bar *CD*.

Solution:

1. Calculate the axial loading on the bar *CD*.

   The bar *CD* is a two-force member. The axial loading P_{CD} can be solved by the resultant moment about point *A*:

   $$\sum M_A = -P \times L_1 + P_{CD} \times \sin(\theta) \times L_2 = 0. \tag{a}$$

   From Equation (a), we have:

   $$P_{CD} = \frac{P \times L_1}{\sin(\theta) \times L_2}. \tag{b}$$

2. The limit state function of the bar *CD*.

   Since the bar material is ductile, we use Equation (4.11) to establish the limit state function:

   $$g\left(S_y, d, L_1, L_2, \theta, P\right) = S_y - \frac{P \times L_1}{\sin(\theta) \times L_2} \frac{1}{(\pi d^2/4)}$$

   $$= S_y - \frac{4P \times L_1}{\pi d^2 L_2 \sin(\theta)} = \begin{cases} > 0 & \text{Safe} \\ = 0 & \text{Limit state} \\ < 0 & \text{Failure.} \end{cases} \tag{c}$$

   There are six random variables in the limit state function (c). All dimension random variables L_1, L_2, d, and θ can be treated as normal distributions. We can use Equation (4.1) to calculate their means and standard deviations. The concentrated loading P can also be treated as a normal distribution. We can use Equation (4.2) to calculate its mean and standard deviation. All distribution parameters of random variables in the limit state function Equation (c) are listed in Table 4.13.

Table 4.13: The distribution parameters of random variables in Equation (c)

S_y (ksi)		d (in)		L_1 (in)		L_2 (in)		θ (radian)		P (klb)	
μ_{S_y}	σ_{S_y}	μ_d	σ_d	μ_{L_1}	σ_{L_1}	μ_{L_2}	σ_{L_2}	μ_θ	σ_θ	μ_P	σ_P
34.5	3.12	0.25	0.00125	20.00	0.0075	41.00	0.0075	0.7854	0.00436	1.90	0.045

3. Reliability R of the bar.

   The limit state function (c) contains six normally distributed random variables and is a non-linear function. We can follow the H-L method discussed in Section 3.6 and the program flowchart in Figure 3.6 to create a MATLAB program. The iterative results are listed in Table 4.14. From the iterative results, the reliability index β and corresponding reliability R of the bar CD in this example are:

$$\beta = 2.4391 \qquad R = \Phi(2.4391) = 0.9926.$$

Table 4.14: The iterative results of Example 4.7 by the H-L method

Iterative#	S_y^*	d^*	L_1^*	L_2^*	θ^*	P^*	β^*	$|\Delta\beta^*|$
1	34.5	0.25	20	41	0.7854	2.4548709	2.4325732	
2	27.113161	0.2496728	20.000074	40.999964	0.7849023	1.9232412	2.4390712	0.006498
3	27.074793	0.2497411	20.000058	40.999972	0.7850065	1.9217741	2.4390644	6.801E-06

∎

Example 4.8

A plate with a centrical hole is subjected to axial loading $P = 2.25 \pm 0.20$ (klb), as shown in Figure 4.4. The diameter of the hole is $d = 0.75 \pm 0.005''$. The height of the plate is $h = 2.50 \pm 0.010''$. The thickness of the plate is $t = 0.25 \pm 0.010''$. The material of the plate is brittle. The ultimate strength S_u follows a two-parameter Weibull distribution with the scale parameter $\eta = 68.5$ (ksi) and the shape parameter $\beta = 1.5$. Determine the reliability of the plate.

Solution:

1. Limit state function of the plate.

   The critical section of this plate will be the section $A-B$, as shown in Figure 4.4, which is the section through the center of the central hole. Since the plate material is brittle, we

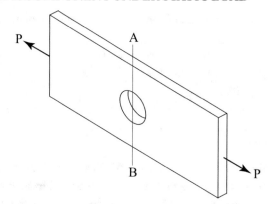

Figure 4.4: A plate with a central hole under an axial loading.

will use Equation (4.12) to establish the limit state function for this plate:

$$g\left(S_u, K_t, d, h, t, P\right) = S_u - K_t \frac{P}{(h-d)\,t} = \begin{cases} > 0 & \text{Safe} \\ = 0 & \text{Limit state} \\ < 0 & \text{Failure.} \end{cases} \qquad \text{(a)}$$

The limit state function contains six random variables. The ultimate strength S_u is a Weibull distribution. All other random variables will be normal distributions. The stress concentration factor K_t in this example has the mean value as 2.36, which can be obtained from current design handbook, or some stress concentration factor websites. Equation (4.10) will be used to calculate the standard deviation of K_t. All dimension random variables d, h, and t can be treated as normal distributions. We can use Equation (4.1) to calculate their means and standard deviations. We can use Equation (4.2) to calculate the mean and the standard deviation of loading P. The distribution parameters of all random variables in the limit state function (a) are listed in Table 4.15.

Table 4.15: The distribution parameters of random variables in Equation (a)

Weibull distribution		Normal distribution									
S_u (ksi)		K_t		d (in)		h (in)		t (in)		P (klb)	
η	β	μ_{K_t}	σ_{K_t}	μ_d	σ_d	μ_h	σ_h	μ_t	σ_t	μ_P	σ_P
68.5	1.5	2.36	0.118	0.75	0.00125	2.50	0.0025	0.25	0.0025	2.25	0.05

2. The reliability of the plate.

The limit state function (a) contains one non-normal distribution random variable. We will use the R-F method to calculate the reliability of the plate, which is discussed in Section 3.7. We can follow the procedure in Section 3.7 and the flowchart in Figure 3.7 to compile a MATLAB program for this example. The iterative resulted are listed in Table 4.16. From the iterative results, the reliability index β and corresponding reliability R of the plate in this example are:

$$\beta = 1.4607 \qquad R = \Phi(1.4607) = 0.9280.$$

Table 4.16: The iterative results of Example 4.8 by the R-F method

Iterative#	S_u^*	K_t^*	d^*	h^*	t^*	P^*	β^*	$|\Delta\beta^*|$
1	61.838053	2.36	0.75	2.5	0.25	11.463622	0.9259472	
2	12.346059	2.3675912	0.7500011	2.499992	0.2499678	2.2810853	1.4606672	0.53472
3	12.17867	2.3665355	0.750001	2.4999931	0.2499722	2.251203	1.4607128	4.558E-05

■

Example 4.9

A circular rod as shown in Figure 4.5 is subjected to an axial loading F_a which is defined by a PMF:

$$P(F_a = F_i) = \begin{cases} p_1 = 0.10 & F_a = F_1 = 2.75\,\text{klb} \\ p_2 = 0.85 & F_a = F_2 = 3.00\,\text{klb} \\ p_3 = 0.05 & F_a = F_3 = 3.25\,\text{klb.} \end{cases} \qquad (a)$$

The diameter of the rod $d = 0.375 \pm 0.005''$. The rod material is ductile. The yield strength S_y of the rod material follows a normal distribution with a mean $\mu_{S_y} = 34.5$ (ksi) and a standard deviation $\sigma_{S_y} = 3.12$ (ksi). Calculate the reliability of the rod.

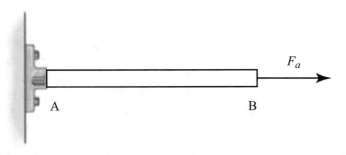

Figure 4.5: A rod under an axial loading.

Solution:

1. The limit state function.

   For this rod made of a ductile material, the limit state function per Equation (4.12) is:

$$g\left(S_y, d, F_a\right) = S_y - \frac{F_a}{\pi d^2/4} = S_y - \frac{4F_a}{\pi d^2} = \begin{cases} > 0 & \text{Safe} \\ 0 & \text{Limit state} \\ < 0 & \text{Failure.} \end{cases} \quad \text{(a)}$$

In this example, F_a is a discrete random variable and described by a PMF. The limit state function of the rod can be expressed as the following three different limit state functions. When $F_a = F_1 = 2.75$ (klb), the limit state function of the rod is:

$$g\left(S_y, d\right) = S_y - \frac{4F_1}{\pi d^2} = \begin{cases} > 0 & \text{Safe} \\ 0 & \text{Limit state} \\ < 0 & \text{Failure.} \end{cases} \quad \text{(b)}$$

When $F_a = F_2 = 3.00$ (klb), the limit state function of the rod is:

$$g\left(S_y, d\right) = S_y - \frac{4F_2}{\pi d^2} = \begin{cases} > 0 & \text{Safe} \\ 0 & \text{Limit state} \\ < 0 & \text{Failure.} \end{cases} \quad \text{(c)}$$

When $F_a = F_3 = 3.25$ (klb), the limit state function of the rod is:

$$g\left(S_y, d\right) = S_y - \frac{4F_3}{\pi d^2} = \begin{cases} > 0 & \text{Safe} \\ 0 & \text{Limit state} \\ < 0 & \text{Failure.} \end{cases} \quad \text{(d)}$$

Per Equation (4.15), the reliability of the rod in this example will be:

$$R = p_1 \times R_1 + p_2 \times R_2 + p_3 \times R_3, \quad \text{(e)}$$

where p_1, p_2, and p_3 are the PMF for the axial loading F_a when it is equal to $F_1 = 2.75$ (klb), $F_2 = 3.00$ (klb) and $F_3 = 3.25$ (klb), respectively. R_1 is the reliability of the rod when $F_a = F_1 = 2.75$ (klb), which is determined by the limit state function (b). R_2 is the reliability of the rod when $F_a = F_2 = 3.00$ (klb), which is determined by the limit state function (c). R_3 is the reliability of the rod when $F_a = F_3 = 3.25$ (klb), which is determined by the limit state function (d).

In this example, the limit state functions (b), (c), and (d) contain only two random variables. The mean and the standard deviation of rod diameter can be calculated by Equation (4.1). The distribution parameters are listed in Table 4.17.

Table 4.17: The distribution parameters of random variables in Equations (b), (c), and (d)

S_y (ksi)		d (in)	
μ_{S_y}	σ_{S_y}	μ_d	σ_d
34.5	3.12	0.375	0.00125

2. Reliability of the rod under axial loading F_a.

The limit state functions (b), (c), and (d) in this example contains two normally distributed random variables and is a nonlinear function. We can follow the H-L method discussed in Section 3.6 and the program flowchart in Figure 3.6 to create a MATLAB program.

The iterative results for the limit state function (b) are listed in the following tables. The reliability index β and corresponding reliability R_1 is:

$$\beta = 3.07292 \qquad R = \Phi(3.07292) = 0.99894.$$

Table 4.18: The iterative results for R_1 by the H-L method for Equation (b)

Iterative#	S_y^*	d^*	β^*	$|\Delta\beta^*|$
1	34.5	0.3185751	3.9023328	
2	22.370303	0.3956266	3.136871	0.7654618
3	24.722992	0.3763318	3.0733175	0.0635535
4	24.924507	0.3748074	3.0729205	0.0003971
5	24.92607	0.3747956	3.0729205	2.357E-08

The iterative results for R_2 of the rod when $F_a = F_2 = 3.0$ (klb) are listed in Table 4.19. From the iterative results, the reliability index β and corresponding reliability R_2 is

$$\beta = 2.3478 \qquad R = \Phi(2.3478) = 0.99056.$$

The iterative results for R_3 of the rod when $F_a = F_3 = 3.25$ (klb) are listed in Table 4.20. From the iterative results, the reliability index β and corresponding reliability R_3 is:

$$\beta = 1.6231 \qquad R = \Phi(1.6231) = 0.94771.$$

Per Equation (e) for this example, the reliability of the rod will be:

$$R = p_1 \times R_1 + p_2 \times R_2 + p_3 \times R_3$$
$$= 0.10 \times 0.99894 + 0.85 \times 0.99056 + 0.05 \times 0.94771 = 0.9893.$$

■

Table 4.19: The iterative results for R_2 by the H-L method for Equation (c)

Iterative#	S_y^*	d^*	β^*	$|\Delta\beta^*|$
1	34.5	0.3327409	2.7990799	
2	25.796855	0.3847976	2.3650156	0.4340643
3	27.131775	0.3752119	2.3478514	0.0171642
4	27.186969	0.3748308	2.3478244	2.702E-05

Table 4.20: The iterative results for R_3 by the H-L method for Equation (d)

Iterative#	S_y^*	d^*	β^*	$|\Delta\beta^*|$
1	34.5	0.3463277	1.8251188	
2	28.823684	0.3788977	1.6262435	0.1988753
3	29.43552	0.3749392	1.6230807	0.0031628
4	29.445978	0.3748726	1.6230798	8.925E-07

4.6.2   RELIABILITY OF A ROD UNDER AXIAL LOADING FOR A DEFORMATION ISSUE

When deformation of a component under axial loading is more than the specified deformation, the component is treated as a failure. The following equation can calculate the deformation of a bar under axial loading:

$$\delta = \int_0^L \frac{F_a(x)\,dx}{EA(x)},\tag{4.17}$$

where $F_a(x)$ is the axial loading function along a rod. $A(x)$ is the rod's cross-section area function along the rod. E is Young's modulus of the rod material. L is the length of the rod under consideration. δ is the axial deformation of the rod segment with a length L. The limit state function of a rod under axial loading for a deformation issue will be:

$$g(E, A, L, F_a) = \Delta - \int_0^L \frac{F_a(x)\,dx}{EA(x)} = \begin{cases} > 0 & \text{Safe} \\ = 0 & \text{Limit state} \\ < 0 & \text{Failure,} \end{cases}\tag{4.18}$$

where Δ is the specified deformation, which is typically treated as a deterministic value. All other variables in Equation (4.18) have the same meanings as those in Equation (4.17).

When a rod can be simplified as several rod segments in each of which axial loading and cross-section area are constant, the deformation of the rod can be expressed as:

$$\delta = \sum \frac{F_{ai} L_i}{E A_i},$$ (4.19)

where F_{ai}, L_i, and A_i are axial loading, length, and cross-section area of the ith segment. Then, the limit state function of a rod under axial loading for a deformation issue will be

$$g(E, A_i, L, F_{ai}) = \Delta - \sum \frac{F_{ai} L_i}{E A_i} = \begin{cases} > 0 & \text{Safe} \\ = 0 & \text{Limit state} \\ < 0 & \text{Failure.} \end{cases}$$ (4.20)

The limit state function of a rod for a deformation issue needs to be determined by the actual problem. For example, there is an allowable envelope dimension L_e, as shown in Figure 4.6. The

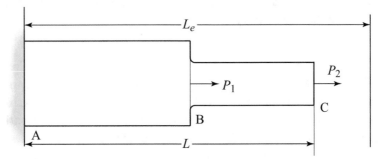

Figure 4.6: Schematic of a rod with an envelope dimension L_e.

design specification could be that the actual physical entire length L of the bar is not allowed to over the envelope dimension L_e, or that a gap must be maintained to be at least Δ between the actual physical entire length L and the envelope dimension L_e. In such a case, the limit state function of a rod under axial loading will be

$$g(L_e, E, A_i, L_i, F_{ai}) = (L_e - L) - \Delta$$

$$= L_e - \sum \left(L_i + \frac{F_{ai} L_i}{E A_i} \right) - \Delta = \begin{cases} > 0 & \text{Safe} \\ 0 & \text{Limit state} \\ < 0 & \text{Failure,} \end{cases}$$ (4.21)

where L_e is the envelope dimension, which could be a random variable or a deterministic value. One example L_e is a distance to another component. Δ is the gap between the actual physical entire length L and the envelope dimension L_e and will be a deterministic value. $\sum \left(L_i + \frac{F_{ai} L_i}{E A_i} \right)$ is the entire physical length L of the rod.

The limit state functions (4.18) or (4.20) or (4.21) can be used to calculate the reliability of a rod under axial loading for a deformation issue.

When axial loading is expressed by a PMF, as shown in Equation (4.14), we can use the total probability theorem Equation (2.24) in Chapter 2 to calculate the reliability of the components. The reliability of a rod under axial loading for a deformation issue will be the same as Equation (4.16) and is shown here again:

$$R = \sum_{i=1}^{n} (p_i \times R_i).$$ \hfill (4.16)

In this case, R_i can be calculated by the limit state functions (4.18), (4.20), or (4.21) when the axial loading F_a is equal to F_{ai}. p_i is the PMF when the axial loading is equal to F_{ai}.

We will use two examples to demonstrate how to calculate the reliability of a rod under axial loading for a deformation issue. Example 4.10 will be the reliability of a rod under axial loading, which is described by a PDF. Example 4.11 will be the reliability of a rod under axial loading, which is described by a PMF.

Example 4.10

A key component: the stepped round-solid bar ABC is subjected to two axial loadings, as shown in Figure 4.7. The bar ABC is fixed at the right end. There is another component (does not show) near the left end of the bar. The dimensions and loadings for this example are listed in Table 4.21. The Young's modulus of the bar material follows a normal distribution with a mean $\mu_E = 2.76 \times 10^7$ (psi) and a standard deviation $\sigma_E = 6.89 \times 10^5$ (psi). The design specification is that the gap Δ between the deformation of the entire bar and the envelope dimension must be at least $0.003''$. Use the Monte Carlo method to calculate the reliability of the bar and its range with a 95% confidence level.

Table 4.21: The dimensions and loading for the Example 4.10

d_1 (in)	d_2 (in)	L_1 (in)	L_2 (in)	L_e	P_1 (lb)	P_2 (lb)
0.750 ± 0.002	0.500 ± 0.002	5.000 ± 0.005	3.000 ± 0.005	8.010 ± 0.005	1300 ± 120	500 ± 50

Solution:

1. Axial loading and geometric parameters in each segment.

   The stepped bar ABC can be divided into two segments. The loading and corresponding geometric parameters for each segment are listed in Table 4.22.

   The deformation of the entire bar per Equation (4.19) will be:

$$\delta = \frac{(P_1 + P_2)L_1}{E\pi d_1^2/4} + \frac{P_2 L_2}{E\pi d_2^2/4} = \frac{4(P_1 + P_2)L_1}{E\pi d_1^2} + \frac{4P_2 L_2}{E\pi d_2^2}.$$ \hfill (a)

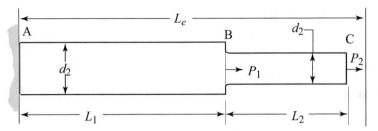

Figure 4.7: A stepped bar with an envelope dimension.

Table 4.22: The loading and corresponding geometric parameters for each segment

	Segment AB			Segment BC	
Length	Cross-section area	Axial loading	Length	Cross-section area	Axial loading
L_1	$\pi d_1^2 / 4$	$P_1 + P_2$	L_2	$\pi d_2^2 / 4$	P_2

2. The limit state function of the stepped bar for the deformation issue.

   The limit state function of this example can be established per Equation (4.20):

   $$g(E, L_e, L_1, L_2, d_1, d_2, P_1, P_2)$$

   $$= L_e - \left(L_1 + L_2 + \frac{4(P_1 + P_2)L_1}{E\pi d_1^2} + \frac{4P_2 L_2}{E\pi d_2^2} \right) - 0.003$$

   $$= \begin{cases} > 0 & \text{Safe} \\ 0 & \text{Limit state} \\ < 0 & \text{Failure.} \end{cases} \qquad (b)$$

   The geometric dimension and the loading can be treated as normal distributions. For geometric dimensions L_e, L_1, L_2, d_1, and d_2, we can use Equation (4.1) to calculate their means and standard deviations. The axial P_1 and P_2 can also be treated as normal distributions. For the loadings P_1 and P_2, we can use Equation (4.2) to calculate their means and standard deviations. All distribution parameters of random variables in the limit state function Equation (b) are listed in Table 4.23.

3. The reliability of the stepped bar.

   The limit state function in this example contains eight normally distributed random variables. We will use the Monte Carlo method to calculate the reliability of this example. The Monte Carlo method has been discussed in Section 3.8. We can follow the Monte Carlo method and the program flowchart in Figure 3.8 to create a MATLAB program. Since the stepped bar is a key component, we will use the trial number $N = 1{,}598{,}400$ from

Table 4.23: The distribution parameters of Example 4.10

E (psi)		L_e (in)		L_1 (in)		L_2 (in)	
μ_E	σ_E	μ_{L_e}	σ_{L_e}	μ_{L_1}	σ_{L_1}	μ_{L_2}	σ_{L_2}
2.76×10^7	6.89×10^5	8.010	0.00125	5.000	0.00125	3.000	0.00125
d_1 (in)		d_2 (in)		P_1 (lb)		P_2 (lb)	
μ_{d_1}	σ_{d_1}	μ_{d_2}	σ_{d_2}	μ_{P_1}	σ_{P_1}	μ_{P_2}	σ_{P_2}
0.750	0.0005	0.500	0.0005	1300	30.0	500	12.5

Table 3.2 in Section 3.8. The estimated reliability of this bar R by the MATLAB program is:

$$R = 0.9971. \tag{c}$$

The estimated probability of the bar failure F is

$$F = 1 - R = 1 - 0.9971 = 0.0029. \tag{d}$$

The relative error of the probability of the failure is

$$\varepsilon = 0.0295. \tag{e}$$

So, the range of the probability of the bar failure with a 95% confidence level will be:

$$F = 0.0029 \pm 0.0029 \times 0.0295 = 0.0029 \pm 0.00009. \tag{f}$$

Therefore, the range of the reliability of the bar with a 95% confidence level will be:

$$R = 1 - F = 0.9971 \pm 0.00009. \tag{g}$$

∎

Example 4.11

A key component: a stepped plate as shown in Figure 4.8 is subjected to an axial loading F_a which is defined by a PMF:

$$P(F_a = F_i) = \begin{cases} p_1 = 0.20 & F_a = F_1 = 3150 \,(\text{lb}) \\ p_2 = 0.75 & F_a = F_2 = 3200 \,(\text{lb}) \\ p_3 = 0.05 & F_a = F_3 = 3250 \,(\text{lb}). \end{cases} \tag{a}$$

The Young's modulus of the plate material follows a normal distribution with a mean $\mu_E = 2.76 \times 10^7$ (psi) and a standard deviation $\sigma_E = 6.89 \times 10^5$ (psi). The design specification is that

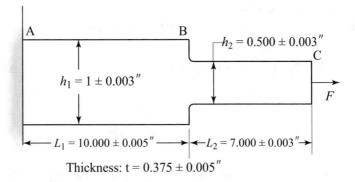

Figure 4.8: A stepped plate under an axial loading.

allowable deformation of the entire stepped plate is less than 0.008″. Use the Monte Carlo method with a 95% confidence level to calculate the reliability of the stepped plate.

Solution:

1. Axial loading and geometric parameters in each segment.

   This stepped-plate can be divided into two segments. The loading and corresponding geometric parameters for each segment are listed in Table 4.24.

Table 4.24: The loading and corresponding geometric parameters for each segment

Segment AB			Segment BC		
Length	Cross-section area	Axial loading	Length	Cross-section area	Axial loading
L_1	$h_1 t$	F	L_2	$h_2 t$	F

The deformation of the entire plate per Equation (4.19) will be:

$$\delta = \frac{FL_1}{Eh_1 t} + \frac{FL_2}{Eh_2 t} = \frac{F(L_1 h_2 + L_2 h_1)}{Eh_1 h_2 t}. \tag{a}$$

2. The limit state function of the stepped plate for the deformation issue.

   The limit state function of this example can be established per Equation (4.20):

$$g(E, L_1, L_2, h_1, h_2, t) = 0.008 - \frac{F(L_1 h_2 + L_2 h_1)}{Eh_1 h_2 t} = \begin{cases} > 0 & \text{Safe} \\ 0 & \text{Limit state} \\ < 0 & \text{Failure.} \end{cases} \tag{b}$$

In this example, the axial loading F is a discrete random variable and described by a PMF. The limit state function of the stepped plate can be expressed as the following three different limit state functions.

When $F = F_1 = 3150$ (lb), the limit state function of the stepped plate is:

$$g\left(E, L_1, L_2, h_1, h_2, t\right) = 0.008 - \frac{3150\left(L_1 h_2 + L_2 h_1\right)}{E h_1 h_2 t} = \begin{cases} > 0 & \text{Safe} \\ 0 & \text{Limit state} \\ < 0 & \text{Failure.} \end{cases} \quad \text{(c)}$$

When $F_a = F_2 = 3200$ (lb), the limit state function of the stepped plate is:

$$g\left(E, L_1, L_2, h_1, h_2, t\right) = 0.008 - \frac{3200\left(L_1 h_2 + L_2 h_1\right)}{E h_1 h_2 t} = \begin{cases} > 0 & \text{Safe} \\ 0 & \text{Limit state} \\ < 0 & \text{Failure.} \end{cases} \quad \text{(d)}$$

When $F_a = F_3 = 3250$ (lb), the limit state function of the stepped plate is:

$$g\left(E, L_1, L_2, h_1, h_2, t\right) = 0.008 - \frac{3250\left(L_1 h_2 + L_2 h_1\right)}{E h_1 h_2 t} = \begin{cases} > 0 & \text{Safe} \\ 0 & \text{Limit state} \\ < 0 & \text{Failure.} \end{cases} \quad \text{(e)}$$

Per Equation (4.16), the reliability of the plate in this example will be:

$$R = p_1 \times R_1 + p_2 \times R_2 + p_3 \times R_3, \quad \text{(f)}$$

where p_1, p_2, and p_3 are the PMF for the axial loading F_a when it is equal to $F_1 = 3150$ (lb), $F_2 = 3200$ (lb), and $F_3 = 3250$ (lb), respectively. R_1 is the reliability of the stepped plate determined by the limit state function (c). R_2 is the reliability of the stepped plate determined by the limit state function (d). R_3 is the reliability of the stepped plate determined by the limit state function (e).

The geometric dimensions and the loading can be treated as normal distributions. For geometric dimensions L_1, L_2, h_1, h_2, and t, we can use Equation (4.1) to calculate their means and standard deviations. All distribution parameters of random variables in the limit state function Equations (c), (d), and (e) are displayed in Table 4.25.

3. The reliability of the stepped plate.

We will use the Monte Carlo simulation method to calculate the reliability R_1, R_2, and R_3 of this example. Then, we can use Equation (f) to calculate the reliability of the stepped plate. Since the stepped bar is a key component, we will use the trial number $N = 1,598,400$ from Table 3.2 in Section 3.8. We can follow the Monte Carlo method

Table 4.25: The distribution parameters of the Example 4.11 for Equations (c), (d), and (e)

E (psi)		L_1 (in)		L_2 (in)		h_1 (in)		h_2 (in)		t (in)	
μ_E	σ_E	μ_{L_1}	σ_{L_1}	μ_{L_2}	σ_{L_2}	μ_{h_1}	σ_{h_1}	μ_{h_2}	σ_{h_2}	μ_t	σ_t
2.76E7	6.89E5	10.000	7.5E-4	7.000	7.5E-4	1.000	7.5E-4	0.50	7.5E-4	0.375	7.5E-4

discussed in Section 3.8 and the program flowchart in Figure 3.8 to create a MATLAB program. The estimated reliability of this component R is:

$$R = 0.9981.$$

■

4.7   RELIABILITY OF A COMPONENT UNDER DIRECT SHEARING

Direct shearing refers that the failure of a component is due to direct shearing stress on the shearing-off surface where the bending moment is small and can be negligible. Figure 4.9a shows a bolted joint. The shearing-off section is shown in Figure 4.9b. In this case, the bending moment on the shearing off section is small and can be negligible. This case can be called as a single shear because there is one shearing off section per shearing-off item.

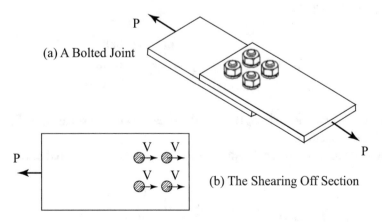

(a) A Bolted Joint

(b) The Shearing Off Section

Figure 4.9: Schematic of a bolted joint.

Another direct shearing example is a joint connection, as shown in Figure 4.10a. Shearing off sections of the pin is shown in Figure 4.10b. In this case, the bending moment on the shearing

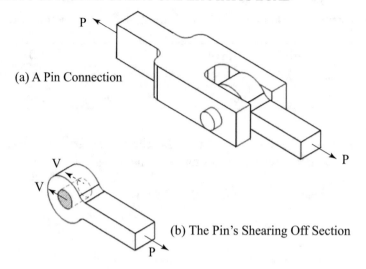

(a) A Pin Connection

(b) The Pin's Shearing Off Section

Figure 4.10: Schematic of a pin connection.

off section is small and can be negligible. The pin has two shearing-off sections in this case. It is also called as double shearing because there are two shearing-off sections per pin item.

For a component under direct shearing, the failure is due to shearing stress. When the shear stress is more than the shearing strength such as the shearing yield strength for ductile material and the ultimate shearing strength for brittle material, the component under direct shearing is treated as a failure. The limit state function of a component made from a ductile material under direct shearing is:

$$g\left(S_{sy}, A, V\right) = S_{sy} - \frac{V}{A} = \begin{cases} > 0 & \text{Safe} \\ = 0 & \text{Limit state} \\ < 0 & \text{Failure,} \end{cases} \qquad (4.22)$$

where S_{sy} is material shear yield strength. V is the shear force on a shearing-off section. A is the area of a shearing-off section.

The limit state function of a component made from a brittle material under direct shearing is:

$$g\left(S_{su}, A, V\right) = S_{su} - \frac{V}{A} = \begin{cases} > 0 & \text{Safe} \\ = 0 & \text{Limit state} \\ < 0 & \text{Failure,} \end{cases} \qquad (4.23)$$

where S_{su} is material shear ultimate strength. V and A have the same meaning as those in Equation (4.22).

When the shear force on a direct shearing section is described by a PMF, we can use the total probability theorem Equation (2.24) in Chapter 2 to calculate the reliability of the components. In this case, the shearing force is expressed by a PMF as:

$$P\left(V = V_i\right) = p_i \qquad\qquad i = 1, 2, \ldots, n, \tag{4.24}$$

where p_i is the PMF when the shearing force V is equal to V_i. Since it is a PMF, the reliability of the component under such the shearing force will be:

$$R = \sum_{i=1}^{n} \left(p_i \times R_i\right), \tag{4.25}$$

where R is the reliability of component under such shearing forces. R_i is the reliability of a component when the direct shearing force V is equal to V_i. Even V_i can be a constant value; we still need to use Equations (4.22) or (4.23) to calculate R_i because other variables in the limit state function are still random variable.

We will use two examples of direct shearing to demonstrate how to calculate the reliability of a component under direct shearing.

Example 4.12

A bar under a tensile loading $P = 0.90 \pm 0.0.08$ (klb) is connected to the ground supporter by a single shearing pin. The pin material is brittle. The ultimate shear strength S_{su} (ksi) of the pin follows a log-normal distribution with the distribution parameters $\mu_{\ln S_{su}} = 3.25$ and $\sigma_{\ln S_{su}} = 0.181$. The pin has a diameter $d = 0.25 \pm 0.005''$. Calculate the reliability of this pin.

Solution:

1. The limit state function of the pin.

    The limit state function of this single shearing pin per Equation (4.22) will be:

$$g\left(S_{su}, d, P\right) = S_{su} - \frac{P}{\pi d^2/4} = S_{su} - \frac{4P}{\pi d^2} = \begin{cases} > 0 & \text{Safe} \\ 0 & \text{Limit state} \\ < 0 & \text{Failure.} \end{cases} \tag{a}$$

    The loading can be simplified as a normal distribution per Equation (4.2). The pin diameter will be treated as a normal distribution per Equation (4.1). The type of distributions and corresponding distribution parameters of random variables in the limit state function (a) are listed in Table 4.26.

2. The reliability of the single shearing pin.

    There are three random variables in the limit state function (a). One is a lognormal distribution, and two are normal distributions. We can follow the procedure of the R-F method in

Table 4.26: The distribution parameters of the limit state function (a) for Example 4.12

Ultimate shear strength S_{su} (ksi) Lognormal distribution		Diameter d (in) Normal distribution		Loading P (klb) Normal distribution	
$\mu_{\ln S_{su}}$	$\sigma_{\ln S_{su}}$	μ_d	σ_d	μ_P	σ_P
3.25	0.181	0.25	0.00125	0.90	0.02

Section 3.7 and the flowchart in Figure 3.7 to compile a MATLAB program. The iterative results are listed in Table 4.27. From the table, the reliability index β and corresponding reliability R of the pin in this example are:

$$\beta = 1.8683 \qquad R = \Phi(1.8683) = 0.9691.$$

Table 4.27: The iterative results of Example 4.12 by the R-F method

Iterative#	S_{su}^*	d^*	P^*	β^*	$|\Delta\beta^*|$
1	26.21628	0.25	1.286889	1.562365	
2	18.41154	0.249893	0.902998	1.868327	0.305963
3	18.44584	0.249872	0.904531	1.868335	7.81E-06

■

Example 4.13
The beam BCD is subjected to a concentrated load $P = 2.01 \pm .20$ (klb), and supported by two vertical bars AB and DE, as shown in Figure 4.11. The bar AB is connected to the supporter through a double shearing pin, as shown in Figure 4.11. The pin material is ductile. Its diameter is $d = 3/16 \pm 0.005''$. The shear yield strength follows a normal distribution with a mean $\mu_{S_{sy}} = 32.2$ (ksi) and a standard deviation $\mu_{S_{sy}} = 3.63$ (ksi). Use the Monte Carlo method to calculate the reliability of the double shearing pin and its range with a 95% confidence level.

Solution:

1. The shearing force on a shearing-off section.

   The loading on the bar AB is $F_{AB} = PL_1/L_2$. Since the pin A is a double shear pin, the shear force on a shearing-off section will be $F_{AB}/2$. So, the shearing force V will be:

   $$V = \frac{F_{AB}}{2} = \frac{PL_1}{2L_2}. \tag{a}$$

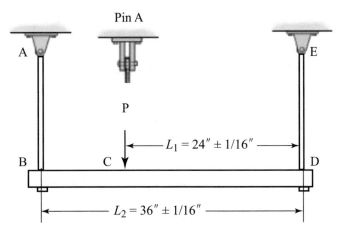

Figure 4.11: A beam with a concentrated load P.

2. Limit state function of the double shear pin A.

   For a pin made of ductile material, its limit state function per Equation (4.22) is:

$$g\left(S_{sy}, d, L_1, L_2, P\right) = S_{sy} - \frac{2PL_1}{\pi L_2 d^2} = \begin{cases} > 0 & \text{Safe} \\ 0 & \text{Limit state} \\ < 0 & \text{Failure.} \end{cases} \quad \text{(b)}$$

   There are five random variables in the limit state function (b). For geometric dimensions, they can be treated as a normal distribution per Equation (4.1). For the loading, it can be treated as a normal distribution per Equation (4.2). All these five random variables are normal distributions. Their distribution parameters are listed in Table 4.28.

Table 4.28: The distribution parameters of the limit state function (b) for Example 4.13

S_{sy} (ksi)		d (in)		L_1 (in)		L_2 (in)		P (klb)	
$\mu_{S_{sy}}$	$\sigma_{S_{sy}}$	μ_d	σ_d	μ_{L_1}	σ_{L_1}	μ_{L_2}	σ_{L_2}	μ_P	σ_P
32.2	3.63	0.1875	0.00125	24	0.015625	36	0.015625	2.01	0.05

3. The reliability of the double shearing pin A.

   We will use the Monte Carlo method to calculate the reliability of this example. The Monte Carlo method has been discussed in Section 3.8. We can follow the Monte Carlo method and the program flowchart in Figure 3.8 to create a MATLAB program. Since the limit state function is not too complicated, we will use the trial number $N = 1,598,400$ from

Table 3.2 in Section 3.8. The estimated reliability of this component R is:

$$R = 0.9840. \tag{c}$$

The estimated probability of the component failure F is

$$F = 1 - R = 1 - 0.9840 = 0.016. \tag{d}$$

The relative error of the probability of the failure is

$$\varepsilon = 0.0124. \tag{e}$$

So, the range of the probability of the component failure with a 95% confidence level will be:

$$F = 0.016 \pm 0.016 \times 0.0124 = 0.016 \pm 0.0002. \tag{f}$$

Therefore, the range of the reliability of the pin with a 95% confidence level will be:

$$R = 1 - F = 0.9840 \pm 0.0002. \tag{g}$$

∎

4.8 RELIABILITY OF A SHAFT UNDER TORSION

4.8.1 RELIABILITY OF A SHAFT UNDER TORSION FOR A STRENGTH ISSUE

A circular shaft under torsion T will have maximum shear stress on the outer layer of the shaft. The following equation will calculate the maximum shear stress:

$$\tau_{\max} = \frac{T d_o/2}{J}, \tag{4.26}$$

where $\tau_{\max}$ is the maximum shear stress on a cross-section. T is resultant internal torque on the cross-section of a shaft. d_o is the outer diameter of a shaft on the cross-section. J is the polar moment inertia of the shaft cross-section.

When the maximum shear stress of a shaft under torsion exceeds shaft material's yield strength, the shaft is treated as a failure. The limit state function of a solid shaft with circular cross-section under torsion will be:

$$g\left(S_{sy}, K_s, d_o, T\right) = S_{sy} - K_s \frac{T d_o/2}{J} = S_{sy} - K_s \frac{T d_o/2}{\pi d_o^4/32} = S_{sy} - K_s \frac{16T}{\pi d_o^3}$$

$$= \begin{cases} > 0 & \text{Safe} \\ = 0 & \text{Limit state} \\ < 0 & \text{Failure,} \end{cases} \tag{4.27}$$

where S_{sy} is material shear yield strength. K_s is the shear stress concentration factor under torsion. T and d_o have the same meaning as those in Equation (4.26).

The limit state function of a hollow shaft with circular cross-section under torsion will be:

$$g\left(S_{sy}, K_s, d_o, d_i, T\right) = S_{sy} - K_s \frac{T d_o/2}{J} = S_{sy} - K_s \frac{T d_o/2}{\pi(d_o^4 - d_i^4)/32}$$

$$= S_{sy} - K_s \frac{16 T d_o}{\pi\left(d_o^4 - d_i^4\right)}$$

$$= \begin{cases} > 0 & \text{Safe} \\ = 0 & \text{Limit state} \\ < 0 & \text{Failure,} \end{cases} \tag{4.28}$$

where d_i is the inner diameter of a hollow shaft. The rests of parameters in Equation (4.28) have the same meaning as those in Equation (4.27).

When the component's torsion is described by a PMF, we can use the total probability theorem Equation (2.24) in Chapter 2 to calculate the reliability of the components. In this case, the torque is expressed by PMF as:

$$P\left(T = T_i\right) = p_i \qquad i = 1, 2, \ldots, n, \tag{4.29}$$

where p_i is the PMF when the torque T is equal to T_i. Since it is a PMF, we have:

$$R = \sum_{i=1}^{n} \left(p_i \times R_i\right), \tag{4.30}$$

where R is the reliability of a shaft under such torsion. R_i is the reliability of a shaft when the torque T is equal to T_i. Even T_i can be a constant value, we still need to use Equations (4.27) or (4.28) to calculate R_i because other variables in the limit state function are still random variables.

Equations (4.27), (4.28), and (4.30) can be used to calculate the reliability of a shaft by the H-L method, R-F method, or Monte Carlo method which have been discussed in Chapter 3.

Two examples will demonstrate how to calculate the reliability of a shaft under torsion.

Example 4.14

The stepped shaft is with a smaller shaft diameter $d_1 = 0.750 \pm 0.005''$ and a larger diameter $d_2 = 1.000 \pm 0.005''$ as shown in Figure 4.12. The fillet radius is $1/16''$. The stepped shaft is subjected to a torque $T = 1000 \pm 160$ (lb/in). The shear yield strength of the shaft material follows a normal distribution with a mean $\mu_{S_{sy}} = 32{,}200$ (psi) and the standard deviation $\sigma_{S_{sy}} = 3630$ (ksi). Use the H-L method to calculate the reliability of the shaft under the torsion.

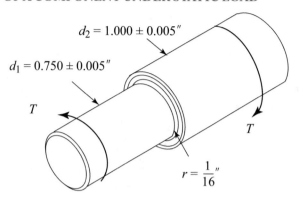

Figure 4.12: A stepped-shaft under a torsion.

Solution:

1. The limit state function of the shaft on the stepped cross-section.

   The critical section for this shaft will be on the stepped section with a fillet because of stress concentration effect. The limit state function of the shaft for this problem is:

   $$g\left(S_{sy}, K_s, d_o, T\right) = S_{sy} - K_s\frac{16T}{\pi d_1^3} = \begin{cases} > 0 & \text{Safe} \\ 0 & \text{Limit state} \\ < 0 & \text{Failure.} \end{cases} \qquad (a)$$

   There are four random variables in the limit state function (a). The shear stress concentration factor K_s can be treated as a normal distribution. Its mean can be obtained from current design handbook or some websites about the stress concentration factor. In this example, the mean of the shear stress concentration factor K_s is 1.92. Use Equation (4.10) to estimate the standard deviations. The shaft diameter $d_1 = 0.750 \pm 0.005''$ can be treated as a normal distribution. Its mean and standard deviation can be calculated by Equation (4.1). The torque $T = 1000 \pm 160$ (lb/in) cab be simplified as a normal distribution. Its mean and standard deviation can be calculated by Equation (4.2). The distribution parameters of all four random variables in the limit state function (a) are listed in Table 4.29.

Table 4.29: The distribution parameters of random variables for Equation (a)

S_{sy} (psi)		K_s		d_1 (in)		T (lb/in)	
Normal distribution		Normal distribution		Normal distribution		Normal distribution	
$\mu_{S_{sy}}$	$\sigma_{S_{sy}}$	μ_{K_s}	σ_{K_s}	μ_{d_1}	σ_{d_1}	μ_T	σ_T
32220	3630	1.92	0.096	0.750	0.00125	1000	40

2. The reliability of the shaft by the H-L method.

The limit state function (a) contains four normally distributed random variables and is a nonlinear function. We will follow the H-L method discussed in Section 3.6 and the program flowchart in Figure 3.6 to create a MATLAB program. The iterative results are listed in Table 4.30. From the iterative results, the reliability index β and corresponding reliability R of the shaft in this example are:

$$\beta = 2.2999 \qquad R = \Phi(2.2999) = 0.9893.$$

Table 4.30: The iterative results of Example 4.14 by the M-L method

Iterative#	S_y^*	K_s^*	d_1^*	T^*	β^*	$|\Delta\beta^*|$
1	32220	1.92	0.75	1390.074	2.215268	
2	25067.94	2.003943	0.749891	1035.755	2.300594	0.085326
3	24537.88	1.987212	0.749909	1022.464	2.299889	0.000705
4	24527.04	1.986438	0.74991	1022.418	2.299888	1.39E-07

■

Example 4.15
A solid shaft with a constant cross-section is subjected to a torque. The diameter of the shaft is $d_o = 1.000 \pm 0.005''$. The torque can be described by a PMF:

$$P(T = T_i) = \begin{cases} 0.45 & T = 3500\,(\text{lb/in}) \\ 0.55 & T = 4500\,(\text{lb/in}). \end{cases}$$

The shear yield strength of the shaft material follows a normal distribution with a mean $\mu_{S_{sy}} = 32{,}200$ (psi) and a standard deviation $\sigma_{S_{sy}} = 3630$ (ksi). Use the Monte Carlo method to calculate the reliability of the shaft.

Solution:

1. Limit state function of the shaft under the torsion.

Since the torque in this example is described by a PMF, we will have two limit state functions for this shaft.

For $T = 3500$ (lb/in),

$$g\left(S_{sy}, d_o\right) = S_{sy} - \frac{16 \times 3500}{\pi d_o^3} = \begin{cases} > 0 & \text{Safe} \\ 0 & \text{Limit state} \\ < 0 & \text{Failure.} \end{cases} \qquad \text{(a)}$$

For $T = 4500$ (lb/in),

$$g\left(S_{sy}, d_o\right) = S_{sy} - \frac{16 \times 4500}{\pi d_o^3} = \begin{cases} > 0 & \text{Safe} \\ 0 & \text{Limit state} \\ < 0 & \text{Failure.} \end{cases} \qquad \text{(b)}$$

There are two random variables in the limit state functions. The shaft diameter $d_o = 1.000 \pm 0.005''$ can be treated as a normal distribution. We can use Equation (4.1) to calculate its mean and standard deviation. Their distribution parameters in this example are listed in Table 4.31.

Table 4.31: The distribution parameters of random variables for Equations (a) and (b)

S_{sy} (psi) Normal distribution		d_0 (in) Normal distribution	
$\mu_{S_{sy}}$	$\sigma_{S_{sy}}$	μ_{d_0}	σ_{d_0}
32220	3630	1.000	0.00125

2. Reliability of the shaft.

The reliability of the shaft in this example can be calculated by Equation (4.30)

$$R = p_1 \times R_1 + p_2 \times R_2, \qquad \text{(c)}$$

where $p_1 = 0.45$ for $T = 2200$ (lb/in). R_1 will be determined by the limit state function (a). $p_2 = 0.55$ for $T = 2500$ (lb/in). R_2 will be determined by the limit state function (b).

We will use the Monte Carlo method to calculate the reliability R_1 and R_2 of this example. The Monte Carlo simulation method has been discussed in Section 3.8. We can follow the Monte Carlo method and the program flowchart in Figure 3.8 to create a MATLAB program. Since this problem is not complicated, we can use the trial number $N = 1,598,400$ for a key component from Table 3.2 in Section 3.8. The estimated reliability of this shaft R is:

$$R = p_1 \times R_1 + p_2 \times R_2 = 0.55 \times 0.99997 + 0.45 \times 0.99466 = 0.99758.$$

∎

4.8.2 RELIABILITY OF A SHAFT UNDER TORSION FOR A DEFORMATION ISSUE

The shaft will be deformed due to an applied torque. The angle of twist between two cross-sections of a shaft is a relative rotational angle ϕ in radian between the two cross-sections. The following equation can calculate the angle of twist for a shaft under torque:

$$\phi = \int_0^L \frac{T(x)\,dx}{G \times J(x)}, \tag{4.31}$$

where ϕ is an angel of twist in radian between the two cross-sections of the shaft with a length L. $T(x)$ is the resultant internal torque at the shaft axial coordinate x. $J(x)$ is the polar moment of inertia of the shaft cross-section at the shaft axial coordinate x. G is the shear Young's modulus.

When a shaft can be divided into several segments each of which is subjected to a constant internal resultant torque and is with a constant cross-section, the angle of twist of such shaft can be expressed as

$$\phi = \sum \frac{T_i L_i}{G J_i}, \tag{4.32}$$

where T_i, L_i, and J_i are the resultant internal torque, the length and the polar moment of inertia in the ith segment. G is the shear Young's modulus. When the angle of twist of a shaft is larger than the specified value, the shaft will be treated as a failure. The limit state function of such shaft under torsion for deformation can be expressed as:

$$g(G, L_i, J_i, T_i) = \Delta - \sum \frac{T_i L_i}{G J_i} = \begin{cases} > 0 & \text{Safe} \\ = 0 & \text{Limit state} \\ < 0 & \text{Failure,} \end{cases} \tag{4.33}$$

where Δ is a specified maximum allowable angle of twist in radian, which is one of the design specifications. Δ is treated as a deterministic value. The rest of the parameters has the same meaning as those in Equation (4.32).

The limit state function (4.33) can be used to calculate the reliability of the shaft under torsion for a deformation issue.

One example will be shown to demonstrate how to calculate the reliability of a shaft under torsion for a deformation issue.

Example 4.16

A schematic of a stepped shaft on an emergency braking is shown in Figure 4.13. The shear Young's modulus follows a normal distribution with a mean $\mu_G = 1.117 \times 10^7$ (psi) and a standard deviation $\sigma_G = 2.793 \times 10^5$ (psi). The diameters for the shaft AB segment and the BCD segment are $d_1 = 0.75 \pm 0.005''$ and $d_2 = 1.25 \pm 0.005''$. The lengths for the segment AB, BC, and CD are $L_1 = 10 \pm 0.010''$, $L_2 = 5 \pm 0.010''$, and $L_3 = 5 \pm 0.010''$. Section D is treated as a fixed end during the emergency braking. The torque at section A is $T_A = 560 \pm 80$ (lb/in). The

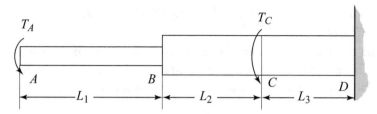

Figure 4.13: The schematic of a shaft.

torque at section C is $T_C = 1380 \pm 120$ (lb/in). The maximum allowable angle of twist for the shaft between section A and the section D is $\Delta = 0.030$ (radian). Use the Monte Carlo method to calculate the reliability of the shaft under torsion for this deformation issue.

Solution:

1. The angle of twist between section A and section D.

   The shaft shown in Figure 4.13 can be divided into three segments: AB segment, BC segment, and CD segment. According to Figure 4.13, the resultant internal torques for the AB segment and the BC segment will be the same and is equal to T_A. The resultant internal torques for the CD segment is $T_A + T_C$. So, the angle of twist of the shaft between the section A and the section D per Equation (4.31) is:

$$\phi = \sum \frac{T_i L_i}{GJ_i} = \frac{T_A L_1}{GJ_{AB}} + \frac{T_A L_2}{GJ_{BC}} + \frac{(T_A + T_C)L_3}{GJ_{CD}}$$
$$= \frac{32 T_A L_1}{G\pi d_1^4} + \frac{32 T_A L_2}{G\pi d_2^4} + \frac{32(T_A + T_C)L_3}{G\pi d_2^4}. \tag{a}$$

2. The limit state function.

   The limit state function of this shaft under torsion per Equation (4.33) is:

$$g(G, d_1, d_2, L_1, L_2, L_3, T_A, T_C) = \Delta - \frac{32 T_A L_1}{G\pi d_1^4} - \frac{32 T_A L_1}{G\pi d_2^4} - \frac{32(T_A + T_C)L_3}{G\pi d_2^4}$$
$$= \begin{cases} > 0 & \text{Safe} \\ 0 & \text{Limit state} \\ < 0 & \text{Failure.} \end{cases} \tag{b}$$

   The limit state function (b) has eight random variables. Geometric dimensions can be treated as normal distributions. Their means and standard deviations can be determined by Equation (4.1). The distribution parameters of eight random variables are listed in Table 4.32.

Table 4.32: The distribution parameters of random variables in Equation (a)

G (psi) Normal distribution		d_1 Normal distribution		d_2 (in) Normal distribution		L_1 (in) Normal distribution	
μ_G	σ_G	μ_{d_1}	σ_{d_1}	μ_{d_2}	σ_{d_2}	μ_{L_1}	σ_{L_1}
1.117×10^7	2.793×10^5	0.75	0.00125	1.000	0.00125	10.000	0.0025
L_2 (in) Normal distribution		L_3 (in) Normal distribution		T_A (lb/in) Normal distribution		T_C (lb/in) Normal distribution	
μ_{L_2}	σ_{L_2}	μ_{L_3}	σ_{L_3}	μ_{T_A}	σ_{T_A}	μ_{T_C}	σ_{T_C}
5.00	0.0025	5.000	0.0025	560	20	1380	30

3. The reliability of the shaft under torsion for a deformation issue.

The Monte Carlo method has been discussed in Section 3.8. We can follow the Monte Carlo method and the program flowchart in Figure 3.8 to create a MATLAB program. Since the simulation problem is not complicated, we can use the trial number $N = 1,598,400$ for a key component from Table 3.2 in Section 3.8. The estimated reliability of this component R is:

$$R = 0.9884.$$

■

4.9 RELIABILITY OF A BEAM UNDER BENDING MOMENT

4.9.1 RELIABILITY OF A BEAM UNDER BENDING FOR A STRENGTH ISSUE

A beam is a long structural element which has a ratio of span to the largest dimension of the cross-section more than ten. A beam primarily resists loads applied laterally to the beam axis. The following equation can calculate the maximum normal stress induced by the bending moment on the critical cross-section:

$$\sigma_{\max} = K_t \frac{Mc}{I} = K_t \frac{M}{I/c} = K_t \frac{M}{Z}, \tag{4.34}$$

where $\sigma_{\max}$ is the maximum bending stress on the critical cross-section of a beam, which could be maximum tensile normal stress or maximum compression normal stress. K_t is stress concentration factor. M is the resultant internal bending moment on the critical cross-section. I is the moment of inertia of the critical cross-section and c is the largest distance between a point on the outer layer of the cross-section and the neutral axis. Z is section modulus of the cross-section and is equal to I/c.

For a beam, the normal stress by bending moment typically is significantly larger than the shear stress induced by the shearing force. Therefore, the shear stress is negligible. When maximum normal stress due to a bending moment in a beam on critical cross-section exceeds material strengths such as the yield strength for ductile material and ultimate tensile strength for brittle material, the beam will be treated as a failure.

For a beam made of brittle material, the limit state function of a beam will be:

$$g\left(S_u, K_t, Z, M\right) = S_u - K_t \frac{M}{Z} = \begin{cases} > 0 & \text{Safe} \\ = 0 & \text{Limit state} \\ < 0 & \text{Failure.} \end{cases} \tag{4.35}$$

For a beam made of ductile material, the limit state function of a beam will be:

$$g\left(S_y, K_t, Z, M\right) = S_y - K_t \frac{M}{Z} = \begin{cases} > 0 & \text{Safe} \\ = 0 & \text{Limit state} \\ < 0 & \text{Failure,} \end{cases} \tag{4.36}$$

where S_u and S_y are the material's ultimate strength and yield strength. The rest of the parameters has the same meaning as those in Equation (4.34).

When a PDF can describe loadings, we can use Equations (4.35) and (4.36) to calculate the reliability of a component by using the H-L method, R-F method, or Monte Carlo method, which has been discussed in Chapter 3.

When loadings on a beam are described by a PMF, we can use the total probability theorem Equation (2.24) in Chapter 2 to calculate the reliability of a beam. The reliability of a beam under such loadings will be:

$$R = \sum_{i=1}^{n} (p_i \times R_i), \tag{4.37}$$

where p_i is the PMF when the bending moment M is equal to M_i. R_i is the reliability of the beam under the bending moment M_i, which can be calculated by using the limit state functions (4.35) or (4.36). R is the reliability of the beam under such bending moments M.

We will show two examples to demonstrate how to use the limit state functions (4.35) and (4.36) to calculate the reliability of a beam under a bending moment.

Example 4.17

A schematic of a beam with a round cross-section is subjected to two concentrated forces P_B and P_D as shown in Figure 4.14. The beam is supported at the sections A and C. The beam is made of a ductile material. The yield strength S_y of the beam's material follows a normal

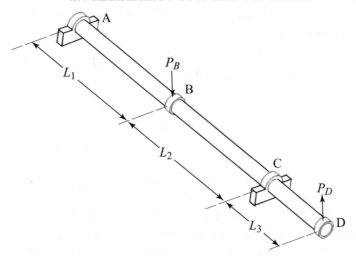

Figure 4.14: A schematic of a beam under two concentrated loads.

distribution with a mean $\mu_{S_y} = 34{,}500$ (psi) and a standard deviation $\sigma_{S_y} = 3120$ (psi). The beam is a round bar with a diameter $d = 1.375 \pm 0.005''$. The concentrated loads at sections B and D are: $P_B = 900 \pm 100$ (lb) and $P_D = 600 \pm 60$ (lb). The geometric dimensions L_1, L_2, and L_3 of the beam are: $L_1 = 10 \pm 0.010''$, $L_2 = 10 \pm 0.010''$, and $L_3 = 5 \pm 0.010''$. Use the H-L method to calculate the reliability of the beam.

1. The maximum bending moment and maximum bending stress.

   According to the shear force and bending moment diagrams of this problem, the maximum bending moment will be in section B and is equal to:

   $$M_B = \frac{P_B L_2 + P_D L_3}{L_1 + L_2} L_1. \tag{a}$$

   Per Equation (4.33), the maximum bending stress at the cross-section B with a maximum bending moment of the beam will be:

   $$\sigma_{\max} = \frac{Mc}{I} = \frac{\frac{P_B L_2 + P_D L_3}{L_1 + L_2} L_1 \times (d/2)}{\pi d^4 / 64} = \frac{32(P_B L_2 + P_D L_3) L_1}{\pi d^3 (L_1 + L_2)}. \tag{b}$$

2. The limit state function of this example.

Since the beam materials is a ductile material, the limit state function of this beam per Equation (4.35) will be:

$$g\left(d, L_1, L_2, L_3, P_B, P_D, S_y\right) = S_y - \frac{32(P_B L_2 + P_D L_3)L_1}{\pi d^3(L_1 + L_2)}$$

$$= \begin{cases} > 0 & \text{Safe} \\ 0 & \text{Limit state} \\ < 0 & \text{Failure.} \end{cases} \qquad \text{(c)}$$

There are seven random variables in the limit state function (c). All dimensional random variables d, L_1, L_2, and L_2 can be treated as normal distributions. We can use Equation (4.1) to calculate their means and standard deviations. The concentrated loading P_B and P_D can also be treated as normal distributions. We can use Equation (4.2) to calculate their means and standard deviations. All distribution parameters of random variables in the limit state function (c) are displayed in Table 4.33.

Table 4.33: The distribution parameters of random variables in Equation (c)

S_{sy} (psi)		d (in)		L_1 (in)		L_2 (in)	
μ_{S_y}	σ_{S_y}	μ_d	σ_d	μ_{L_1}	σ_{L_1}	μ_{L_2}	σ_{L_2}
34500	3120	1.375	0.00125	10	0.0025	10	0.0025
L_3 (in)		P_B (lb)		P_D (lb)			
μ_{L_3}	σ_{L_3}	μ_{P_B}	σ_{P_B}	μ_{P_D}	σ_{P_D}		
5	0.0025	900	25	600	15		

3. Reliability R of the beam under bending.

The limit state function (c) contains seven normally distributed random variables and is a nonlinear function. We will follow the H-L method discussed in Section 3.6 and the program flowchart in Figure 3.6 to create a MATLAB program. The iterative results are listed in Table 4.34. From the iterative results, the reliability index β and corresponding reliability R of the beam in this example are:

$$\beta = 2.7810 \qquad R = \Phi(2.7810) = 0.9973.$$

∎

Table 4.34: The iterative results of Example 4.17 by the H-L method

#	d^*	L_1^*	L_2^*	L_3^*	P_B^*	P_D^*	S_y^*	β^*	$|\Delta\beta^*|$
1	1.4375	10	10	5	900	600	25718.04	2.785902	
2	1.437448	10.00001	10	5.000006	909.4704	601.7047	25912.47	2.780998	0.004903
3	1.437447	10.00001	10	5.000006	909.4547	601.7018	25912.17	2.781004	5.89E-06

Example 4.18

A schematic of a segment of a beam is shown in Figure 4.15. A PMF can describe the internal bending moments on the stepped section.

$$P\left(M = M_i\right) = \begin{cases} p_1 = 0.72 & M_i = 1600 \,(\text{lb/in}) \\ p_2 = 0.28 & M_i = 2000 \,(\text{lb/in}). \end{cases} \tag{a}$$

The dimensions of the stepped cross-section are the heights $h_1 = 1 \pm 0.005''$, $h_2 = 1.5 \pm 0.005''$ and the thickness $b = 0.5 \pm 0.005''$. The radius of the fillet is $r = 1/8''$. The beam material is brittle. The ultimate material strength S_u (ksi) follows a lognormal distribution with a log mean $\mu_{\ln S_u} = 4.10$ and a standard deviation $\mu_{\ln S_u} = 0.196$. Calculate the reliability of the stepped beam.

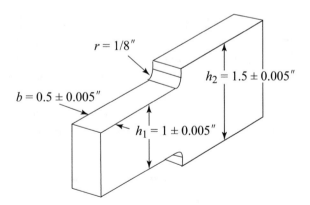

Figure 4.15: A schematic of a segment beam.

Solution:

1. The maximum bending stress of the beam.

The maximum bending stress of this beam will be on the stepped cross-section and can be calculated per Equation (4.34):

$$\sigma_{\max} = K_t \frac{Mc}{I} = K_t \frac{M(h_1/2)}{bh_1^3/12} = \frac{6K_t M}{bh_1^2}. \tag{b}$$

2. The limit state functions of the beam.

Since the beam material is brittle, we will use Equation (4.35) to establish the limit state function.

When the resultant internal bending moment M is equal to $M_1 = 1.6$ (klb.in), the limit state function of the beam is:

$$g\,(S_u, K_t, b, h_1) = S_u - \frac{6K_t M_1}{bh_1^2} = \begin{cases} > 0 & \text{Safe} \\ 0 & \text{Limit state} \\ < 0 & \text{Failure.} \end{cases} \tag{c}$$

When the internal resultant bending moment M is equal to $M_2 = 2.0$ (klb.in), the limit state function of the beam is:

$$g\,(S_u, K_t, b, h_1) = S_u - \frac{6K_t M_2}{bh_1^2} = \begin{cases} > 0 & \text{Safe} \\ 0 & \text{Limit state} \\ < 0 & \text{Failure.} \end{cases} \tag{d}$$

There are four random variables in the limit state functions (c) and (d). The stress concentration factor K_t can be treated as a normal distribution. Its mean can be obtained from design handbook by using the nominal dimensions, that is, the heights $h_1 = 1''$, $h_2 = 1.5''$ and the radius of the fillet $r = 1/8''$. The stress concentration factor under bending for this example is 1.72. Its standard deviation can be estimated per Equation (4.10). The geometric dimensions can be treated as normal distributions per Equation (4.1). The distribution parameters of four random variables in the limit state functions (c) and (d) are listed in Table 4.35.

Table 4.35: The distribution parameters of random variables in Equations (c) and (d)

S_u (psi)		K_t (in)		b (in)		h_1 (in)	
$\mu_{\ln S_u}$	$\sigma_{\ln S_u}$	μ_{K_t}	σ_{K_t}	μ_b	σ_b	μ_{h_1}	σ_{h_1}
4.10	0.196	1.72	0.086	0.5	0.00125	1	0.00125

3. The reliability of the beam.

The reliability of the beam in this example will be calculated per Equation (4.37) because the internal bending moments are specified by a PMF:

$$R = p_1 \times R_1 + p_2 \times R_2 = 0.72R_1 + 0.28R_2, \quad (e)$$

where R_1 is determined by the limit state function (c) and R_2 by the limit state function (d). The limit state functions (c) and (d) contains one lognormal distribution random variable and three normal distributions. We can follow the procedure of the R-F method in Section 3.7 and the flowchart in Figure 3.7 to create a MATLAB program. The iterative results for the limit state function (c) are listed in Table 4.36. The reliability index β and corresponding reliability R_1 of the beam in this example are:

$$\beta = 2.9826 \qquad R_1 = \Phi(2.9826) = 0.9986.$$

Table 4.36: The iterative results of Example 4.18 for the limit state function (c)

Iterative #	S_u^*	K_t^*	b^*	h_1^*	β^*	$|\Delta\beta^*|$
1	61.51051	1.72	0.5	0.732724	3.510922	
2	19.32361	1.794619	0.499946	1.335415	3.359916	0.151006
3	28.96702	1.788618	0.499948	1.08888	3.017535	0.342381
4	33.58777	1.781819	0.499953	1.009281	2.983043	0.034492
5	34.19932	1.781332	0.499954	1.000079	2.982618	0.000425
6	34.20802	1.781339	0.499954	0.999954	2.982618	7.77E-08

The iterative results for the limit state function (d) are listed in Table 4.37. The reliability index β and corresponding reliability R_2 of the beam in this example are:

$$\beta = 1.87774 \qquad R_2 = \Phi(2.9826) = 0.9698.$$

Therefore, the reliability of the beam in this example per Equation (e) is

$$R = 0.72R_1 + 0.28R_2 = 0.9905.$$

■

Table 4.37: The iterative results of Example 4.18 for the limit state function (d)

Iterative #	S_u^*	K_t^*	b^*	h_1^*	β^*	$|\Delta\beta^*|$
1	61.51051	1.72	0.5	0.81921	2.086685	
2	35.95712	1.76435	0.499968	1.085224	1.909968	0.176717
3	41.5102	1.759636	0.49997	1.008678	1.87811	0.031858
4	42.21348	1.759073	0.499971	1.00008	1.877739	0.000371
5	42.22275	1.759077	0.499971	0.999971	1.877739	5.84E-08

4.9.2   RELIABILITY OF A BEAM UNDER BENDING FOR A DEFLECTION ISSUE

When the maximum deflection of a beam exceeds the allowable deflection, the beam is treated as a failure. The general limit state function of a beam under bending for a deflection is:

$$g(y_{max}) = \Delta - y_{max} = \begin{cases} > 0 & \text{Safe} \\ = 0 & \text{Limit state} \\ < 0 & \text{Failure,} \end{cases} \tag{4.38}$$

where Δ is the allowable deflection and is typically treated as a deterministic constant. y_{max} is the maximum deflection of a beam and is a function of other random variables.

There is no general formula for the maximum deflection of a beam. It will be determined per case. For a simple support beam under a concentrated load at the middle of the beam, as shown in Figure 4.16a, the limit state function for a deflection issue is:

$$g(E, I, L, P) = \Delta - y_{max} = \Delta - \frac{PL^3}{48EI} = \begin{cases} > 0 & \text{Safe} \\ 0 & \text{Limit state} \\ < 0 & \text{Failure,} \end{cases} \tag{4.39}$$

where E is the beam's material Young's modulus, I is the moment of inertia of the cross-section with a maximum deflection, P is the external concentrate forces on the middle of the beam, and L is the beam length. Typically, E, I, L, and P are random variables.

For a cantilever beam under a concentrated load at the free end of the beam, as shown in Figure 4.16b, the limit state function for a deflection issue is:

$$g(E, I, L, P) = \Delta - y_{max} = \Delta - \frac{PL^3}{3EI} = \begin{cases} > 0 & \text{Safe} \\ 0 & \text{Limit state} \\ < 0 & \text{Failure,} \end{cases} \tag{4.40}$$

where P is a concentrated force at the free end of a cantilever beam. The rests of the symbols have the same meaning as those in Equation (4.39).

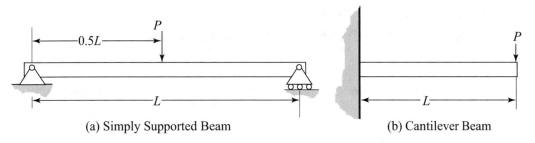

(a) Simply Supported Beam   (b) Cantilever Beam

Figure 4.16: Two typical beams.

We will use one example to demonstrate how to calculate the reliability of a beam for a deflection issue.

Example 4.19

A simple support beam as shown in Figure 4.17 is subjected to a uniform distributed load $w = 10 \pm 1.5$ (lb/in). The span of the beam is $L = 20 \pm 0.065''$. The beam has a constant rectangular cross-section with the height $h = 1.25 \pm 0.005''$ and the thickness $b = 0.5 \pm 0.005''$. The Young's modulus of the beam material follows a normal distribution with a mean $\mu_E = 2.76 \times 10^7$ (psi) and a standard deviation $\sigma_E = 6.89 \times 10^5$ (psi). If the maximum allowable deflection of the beam is $\Delta = 0.010''$. Calculate the reliability of this beam for a deformation issue.

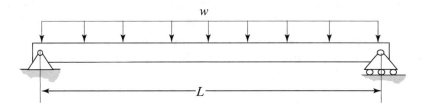

Figure 4.17: A simple support beam under a uniform distributed load w.

Solution:

1. The maximum deflection of the beam under a uniform distributed load.

   For a simple support beam under a uniform distributed load, the maximum deflection of the beam will happen in the middle of the beam and can be calculated by the following equation:

$$y_{max} = \frac{5wL^4}{384EI} = \frac{5wL^4}{384E(bh^3/12)} = \frac{5wL^4}{32Ebh^3}, \tag{a}$$

   where E is material Young's modulus, w is the uniform distributed load, L is the span of the beam, and h and b are the height and thickness of the beam.

2. The limit state function of the beam for a deformation issue.

The limit state function of this beam per Equation (4.38) will be:

$$g(E, w, b, h, L) = \Delta - y_{max} = 0.010 - \frac{5wL^4}{32Ebh^3} = \begin{cases} > 0 & \text{Safe} \\ 0 & \text{Limit state} \\ < 0 & \text{Failure.} \end{cases} \quad (b)$$

The uniform distributed load can be treated as a normal distribution per Equation (4.2). The geometric dimensions can be treated as normal distributions per Equation (4.1). There are five normal distributed random variables in the limit state function (b). Their distribution parameters are listed in Table 4.38.

Table 4.38: The distribution parameters of random variables in Equation (b)

E (psi)		w (lb/in)		b (in)		h (in)		L (in)	
μ_E	σ_E	μ_w	σ_w	μ_b	σ_b	μ_h	σ_h	μ_L	σ_L
2.76×10^7	6.89×10^5	10	0.375	0.5	0.00125	1.25	0.00125	20	0.01625

3. The reliability of the beam for a deformation issue.

The limit state function (b) is a nonlinear equation with five normal distributed random variable. We can follow the H-L method discussed in Section 3.6 and the program flowchart in Figure 3.6 to create a MATLAB program. The iterative results are listed in Table 4.39. From the iterative results, the reliability index β and corresponding reliability R of this beam are:

$$\beta = 1.68155 \qquad R = \Phi(1.68155) = 0.9537.$$

Table 4.39: The iterative results of Example 4.19 by the H-L method

#	E^*	w^*	b^*	h^*	L^*	β^*	$|\Delta\beta^*|$
1	27,600,000	10	0.5	1.25	20.37968	1.641913	
2	26,977,766	10.50873	0.499887	1.24983	20.01083	1.681462	0.039549
3	26,931,076	10.50869	0.499881	1.249822	20.00203	1.681553	9.11E-05

■

4.10   RELIABILITY OF A COMPONENT UNDER COMBINED STRESSES

4.10.1   RELIABILITY OF A COMPONENT OF DUCTILE MATERIAL UNDER COMBINED STRESSES

The maximum-shear-stress (MSS) theory and the distortion-energy (DE) theory are two widely accepted failure theory for a component of a ductile material when it is subjected to combined stresses due to static loadings [5]. These failure theories can be used to establish the limit state function of a component under combined stresses due to static loading.

The MSS theory predicts that yielding begins whenever the maximum shear stress in any element equals or exceeds the maximum shear stress in a tensile-test specimen of the same material when that specimen begins to yield. According to the MSS theory for a ductile material, the limit state function of a component under combined stress due to the static loadings is:

$$g\left(S_y, \tau_{\max}\right) = \frac{S_y}{2} - \tau_{\max} = \begin{cases} > 0 & \text{Safe} \\ = 0 & \text{Limit state} \\ < 0 & \text{Failure,} \end{cases} \tag{4.41}$$

where S_y is the yield strength and $\tau_{\max}$ is the maximum shear stress at the critical point. $\tau_{\max}$ will be determined per case and will be the functions of static loading and component geometric dimensions.

If three principal stress σ_1, σ_2, and σ_3 at the critical point are known and are arranged in such a way $\sigma_1 \geq \sigma_2 \geq \sigma_3$, the maximum shear stress in such case will be:

$$\tau_{\max} = \frac{(\sigma_1 - \sigma_3)}{2}. \tag{4.42}$$

If the stress status of a component at the critical point is plane stress and when two normal stress σ_x, σ_y and shear stress τ_{xy} are known, the two principal stress σ_A and σ_B can be calculated and are arranged in such a way $\sigma_A \geq \sigma_B$:

$$\sigma_A = \frac{\sigma_x + \sigma_y}{2} + \sqrt{\left(\frac{\sigma_x - \sigma_y}{2}\right)^2 + \tau_{xy}^2}$$

$$\sigma_B = \frac{\sigma_x + \sigma_y}{2} - \sqrt{\left(\frac{\sigma_x - \sigma_y}{2}\right)^2 + \tau_{xy}^2}. \tag{4.43}$$

For a plane stress, σ_A and σ_B are two principal stresses. Another principal stress is 0. The maximum shear stress for such case will be:

$$\tau_{\max} = \begin{cases} \dfrac{\sigma_A}{2} & \text{when } \sigma_A \geq \sigma_B \geq 0 \\ \dfrac{\sigma_A - \sigma_B}{2} & \text{when } \sigma_A \geq 0 \geq \sigma_B \\ -\dfrac{\sigma_B}{2} & \text{when } 0 \geq \sigma_A \geq \sigma_B. \end{cases} \tag{4.44}$$

The DE theory predicts that yielding occurs when the distortion strain energy per unit volume reaches or exceeds the distortion strain energy per unit volume for yield in simple tension or compression of the same material. According to the DE theory for a ductile material, the limit state function of a component under combined stresses due to static loadings is:

$$g\left(S_y, \sigma_{von}\right) = S_y - \sigma_{von} = \begin{cases} > 0 & \text{Safe} \\ = 0 & \text{Limit state} \\ < 0 & \text{Failure,} \end{cases} \tag{4.45}$$

where S_y is the yield strength and σ_{von} is the Von-Mises stress of the component at the critical point. σ_{von} will be determined per case and will be the function of static loadings and component geometric dimensions

If three principal stresses σ_1, σ_2, and σ_3 at the critical point are known, the Von-Mises stress σ_{von} is:

$$\sigma_{von} = \sqrt{\frac{(\sigma_1 - \sigma_2)^2 + (\sigma_2 - \sigma_3)^2 + (\sigma_3 - \sigma_1)^2}{2}}. \tag{4.46}$$

If the stress status of the critical point is known, the Von-Mises stress σ_{von} is:

$$\sigma_{von} = \sqrt{\frac{\left(\sigma_x - \sigma_y\right)^2 + \left(\sigma_y - \sigma_z\right)^2 + (\sigma_z - \sigma_x)^2 + 6(\tau_{xy}^2 + \tau_{yz}^2 + \tau_{zx}^2)}{2}}. \tag{4.47}$$

Limit state functions (4.41) and (4.45) can be used to calculate the reliability of a component of a ductile material under combined stresses due to static loading.

We will use three examples to demonstrate how to calculate the reliability of components of a ductile material at the critical point under combined stress due to static loadings.

Example 4.20

A segment of a solid shaft with a diameter $d = 1.125 \pm 0.005''$ is shown in Figure 4.18. The resultant internal torsion and the resultant internal bending moment of a shaft at the critical section are $T = 1400 \pm 120$ (lb/in), and $M = 3500 \pm 450$ (lb/in). The yield strength S_y of the shaft's material follows a normal distribution with the mean $\mu_{S_y} = 34{,}500$ (ksi) and the standard deviation $\sigma_{S_y} = 3120$ (ksi). Calculate the reliability of the shaft by using the DE theory.

Solution:

1. The Von-Mises stress.

   As shown in Figure 4.18, the critical points on this critical section are point A and point B because there are the maximum values of bending stress. Stress elements at points A and B are shown in Figure 4.18b and c, where σ_M and τ_T are the bending stress due to the bending moment M and the shear stress due to the torque T, respectively. Per Equation (4.46), the

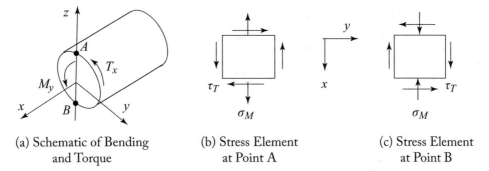

(a) Schematic of Bending
and Torque

(b) Stress Element
at Point A

(c) Stress Element
at Point B

Figure 4.18: Schematic of a segment of a shaft under combined stress.

Von-Mises stress at points A and B in this example are the same. The point A is used as an example to run the calculation. At point A, we have:

$$\sigma_x = \sigma_M = \frac{32M_y}{\pi d^3}, \; \sigma_y = \sigma_z = 0, \quad \tau_{xy} = \tau_T = -\frac{16T_x}{\pi d^3}, \quad \tau_{yz} = \tau_{zx} = 0. \qquad (a)$$

Per Equation (4.46), we have:

$$\sigma_{von} = \sqrt{\left(\frac{32M_y}{\pi d^3}\right)^2 + 3\left(-\frac{16T_x}{\pi d^3}\right)^2} = \frac{16}{\pi d^3}\sqrt{4M_y^2 + 3T_x^2}. \qquad (b)$$

2. The limit state function.

By using the DE theory, the limit state function of this example per Equation (4.44) is:

$$g\left(S_y, d, T_x, M_y\right) = S_y - \frac{16}{\pi d^3}\sqrt{4M_y^2 + 3T_x^2} = \begin{cases} > 0 & \text{Safe} \\ 0 & \text{Limit state} \\ < 0 & \text{Failure.} \end{cases} \qquad (c)$$

There are four random variables in this limit state function. The diameter d can be treated as a normal distribution. Its distribution parameters can be determined per Equation (4.1). The bending moment and the torque will be treated as normal distributions. Their distribution parameters can be determined per Equation (4.2). Their distribution parameters are listed in Table 4.40.

3. The reliability of the shaft under combined stress.

The limit state function (c) is a nonlinear equation with four normal distributed random variable. We can follow the H-L method discussed in Section 3.6 and the program flowchart in Figure 3.6 to create a MATLAB program. The iterative results are listed in

Table 4.40: The distribution parameters of random variables in Equation (c)

S_y (psi)		d (in)		T_x (lb/in)		M_y (lb/in)	
μ_{S_y}	σ_{S_y}	μ_d	σ_d	μ_{T_x}	σ_{T_x}	μ_{M_y}	σ_{M_y}
34500	3120	1.125	0.00125	1400	30	3500	90

Table 4.41. From the iterative results, the reliability index β and corresponding reliability R of this shaft are:

$$\beta = 2.51570 \qquad R = \Phi(2.51570) = 0.99406.$$

Table 4.41: The iterative results of Example 4.20 by the H-L method

Iterative #	S_y^*	d^*	T_x^*	M_y^*	β^*	$|\Delta\beta^*|$
1	34500	1.125	1400	4667.651	2.539208	
2	26737.03	1.124885	1401.118	3533.743	2.515704	0.023504
3	26800.86	1.124912	1401.433	3543.36	2.515703	1.04E-06

■

Example 4.21

Schematic of a thin-cylindrical vessel is depicted in Figure 4.19. The vessel has an inner diameter $d = 48'' \pm 0.125''$ and a thickness $t = 3/8'' \pm 0.060''$. The internal pressure is $p = 350 \pm 30$ (psi). The vessel material is ductile. The yield strength S_y of this material follow a normal distribution with the mean $\mu_{S_y} = 34500$ (psi) and the standard deviation $\sigma_{S_y} = 3120$ (psi). Calculate the reliability of this vessel by using the MSS theory.

Solution:

1. The maximum shear stress.

   Stress element of this vessel at the critical point is shown in Figure 4.19b, where p is the internal pressure, σ_l is the longitudinal normal stress and σ_h is the normal stress in the hoop direction. σ_l and σ_h can be calculated by using the following equation:

   $$\sigma_h = \frac{pd}{2t}, \qquad \sigma_l = \frac{pd}{4t}, \qquad \text{(a)}$$

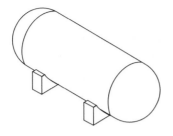

(a) Schematic of a Thin-Wall Vessel

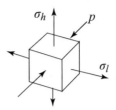

(b) Stress Element of a Critical Point

Figure 4.19: Schematic of a thin-wall cylindrical vessel.

where p is the internal pressure. d is the inner diameter of the vessel, and t is the wall thickness.

Since σ_h, σ_l, and $-p$ are three principal stresses and are arranged as $\sigma_h > \sigma_l > -p$ in this case, the maximum shear stress per Equation (4.41) will be:

$$\tau_{\max} = \frac{\sigma_h + p}{2} = p\left(\frac{d}{4t} + \frac{1}{2}\right). \tag{b}$$

2. The limit state function of the thin-cylindrical vessel.

Per Equation (4.41), the limit state function of this vessel by using the maximum shear stress theory is

$$g\left(S_y, t, p, d\right) = \frac{S_y}{2} - p\left(\frac{d}{4t} + \frac{1}{2}\right) = \begin{cases} > 0 & \text{Safe} \\ 0 & \text{Limit state} \\ < 0 & \text{Failure.} \end{cases} \tag{c}$$

There are four random variables in this limit state function. The geometric dimensions d and t can be treated as a normal distribution. Their distribution parameters can be determined per Equation (4.1). The internal pressure p can be treated as a normal distribution too. Its distribution parameter can be determined per Equation (4.2). Their distribution parameters are listed in Table 4.42.

Table 4.42: The distribution parameters of random variables in Equation (c)

S_y (psi)		t (in)		p (psi)		d (in)	
μ_{S_y}	σ_{S_y}	μ_t	σ_t	μ_p	σ_p	μ_d	σ_d
34500	3120	0.375	0.015	350	7.5	48	0.03125

3. Reliability of the thin-cylindrical vessel under combined stress.

The limit state function (c) is a nonlinear equation with four normal distributed random variable. We can follow the H-L method discussed in Section 3.6 and the program flowchart in Figure 3.6 to create a MATLAB program. The iterative results are listed in Table 4.43. The reliability index β and corresponding reliability R of this vessel are:

$$\beta = 3.56441 \qquad R = \Phi\,(3.56441) = 0.99982.$$

Table 4.43: The iterative results of Example 4.21 by the H-L method

Iterative #	S_y^*	t^*	p^*	d^*	β^*	$|\Delta\beta^*|$
1	34500	0.375	350	73.17857	3.371331	
2	25083.84	0.35518	355.3634	49.4314	3.568117	0.196786
3	24077.35	0.358227	354.2512	47.97873	3.56444	0.003677
4	24029.32	0.358973	354.1123	48.00033	3.564409	3.12E-05

∎

Example 4.22
A plane stress element of a component at the critical point is shown in Figure 4.20. The normal stress σ_x (ksi), the normal compression stress σ_y (ksi), and the shear stress τ_{xy} (ksi) follows normal distributions. Their distribution parameters are listed in Table 4.44.

Table 4.44: The distribution parameters of a plane stress element

σ_x (ksi)		σ_y (ksi)		τ_{xy} (ksi)	
μ_{σ_x}	σ_{σ_x}	μ_{σ_y}	σ_{σ_y}	$\mu_{\tau_{xy}}$	$\sigma_{\tau_{xy}}$
15.2	1.12	17.5	2.1	10.5	0.98

The yield strength S_y of the shaft's material follows a normal distribution with a mean $\mu_{S_y} = 50.5$ (ksi) and a standard deviation $\sigma_{S_y} = 4.12$ (ksi). Calculate the reliability of this component by using the MSS theory.

1. The maximum shear stress.

As shown in Figure 4.20, σ_x is normal tensile stress, σ_y is normal compressive stress, and τ_{xy} is positive shear stress. For this plane stress case, σ_A and σ_B can be calculated per Equa-

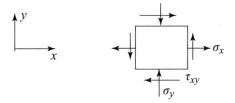

Figure 4.20: **A plane stress element.**

tion (4.43).

$$\sigma_A = \frac{\sigma_x - \sigma_y}{2} + \sqrt{\left(\frac{\sigma_x + \sigma_y}{2}\right)^2 + \tau_{xy}^2}$$

$$\sigma_B = \frac{\sigma_x - \sigma_y}{2} - \sqrt{\left(\frac{\sigma_x + \sigma_y}{2}\right)^2 + \tau_{xy}^2}.$$

(a)

From Equation (a), it is clear that $\sigma_A > 0$ and $\sigma_B < 0$. Therefore, the maximum shear stress in this case per Equation (4.44) is

$$\tau_{\max} = \frac{\sigma_A - \sigma_B}{2} = \sqrt{\left(\frac{\sigma_x + \sigma_y}{2}\right)^2 + \tau_{xy}^2}.$$

(b)

2. The limit state function of this example.

Per Equation (4.41), the limit state function of this example by using the MSS theory is:

$$g\left(S_y, \sigma_x, \sigma_y, \tau_{xy}\right) = \frac{S_y}{2} - \sqrt{\left(\frac{\sigma_x + \sigma_y}{2}\right)^2 + \tau_{xy}^2}.$$

(c)

There are four normally distributed random variables in Equation (c). Their distribution parameters are listed in Table 4.45.

Table 4.45: **The distribution parameters of random variables in Equation (c)**

S_y (ksi)		σ_x (ksi)		σ_y (ksi)		τ_{xy} (ksi)	
μ_{S_y}	σ_{S_y}	μ_{σ_x}	σ_{σ_x}	μ_{σ_y}	σ_{σ_y}	$\mu_{\tau_{xy}}$	$\sigma_{\tau_{xy}}$
50.5	4.12	15.2	1.12	17.5	2.1	10.5	0.98

3. Reliability of the component in this example.

The limit state function (c) is a nonlinear equation with four normal distributed random variables. We can follow the H-L method discussed in Section 3.6 and the program flowchart in Figure 3.6 to create a MATLAB program. The iterative results are listed in Table 4.46. The reliability index β and corresponding reliability R of this component are:

$$\beta = 2.65702 \qquad R = \Phi\,(2.65702) = 0.99606.$$

Table 4.46: The iterative results of Example 4.22 by the H-L method

Iterative #	S_y^*	σ_x^*	σ_y^*	τ_{xy}^*	β^*	$|\Delta\beta^*|$
1	50.5	15.2	17.5	19.24162	2.885932	
2	42.52771	16.21553	21.07022	10.22716	2.642539	0.243393
3	43.32856	16.22218	21.09359	11.0102	2.656992	0.014453
4	43.28253	16.22313	21.09694	10.96119	2.656799	0.000193
5	43.28351	16.22336	21.09776	10.96125	2.656979	0.00018
6	43.28302	16.22343	21.09801	10.96049	2.65702	4.04E-05

4.10.2 RELIABILITY OF A COMPONENT OF BRITTLE MATERIAL UNDER COMBINED STRESSES

The maximum normal stress (MNS) theory and the Brittle Coulomb–Mohr (BCM) theory are two failure theory for a component of brittle material when it is subjected to combined stresses due to static loadings [5].

The MNS theory states that failure occurs whenever one of the three principal stress equals to or exceeds the strength. σ_1, σ_2, and σ_3 are three principal stresses at the critical point which are arranged in such a way $\sigma_1 \geq \sigma_2 \geq \sigma_3$. According to the MNS theory for brittle materials, the limit state functions of a component under combined stresses due to static loadings are as follows.

When $\sigma_1 \geq \sigma_2 \geq \sigma_3 \geq 0$,

$$g\,(S_{ut}, \sigma_1) = S_{ut} - \sigma_1 = \begin{cases} > 0 & \text{Safe} \\ = 0 & \text{Limit state} \\ < 0 & \text{Failure,} \end{cases} \tag{4.48}$$

where S_{ut} is the ultimate tensile strength of brittle material. σ_1 will be determined per case and is the function of loading and component geometric dimensions.

When $0 \geq \sigma_1 \geq \sigma_2 \geq \sigma_3$,

$$g\left(S_{uc}, \sigma_3\right) = S_{ut} - |\sigma_3| = \begin{cases} > 0 & \text{Safe} \\ = 0 & \text{Limit state} \\ < 0 & \text{Failure,} \end{cases} \tag{4.49}$$

where S_{uc} is the ultimate compression strength of brittle material. σ_3 will be determined per case and is the function of loading and component geometric dimensions.

The MNS theory is not recommended to be used when the principal stresses at the critical point have both tensile stress and compressive stress.

BCM theory is a failure theory for a component of brittle material under plane stress due to static loading. Let σ_A and σ_B represent two principal stresses of a component of brittle material at the critical point under a plane stress condition. Let also assume $\sigma_A \geq \sigma_B$. According to the BCM theory and the different conditions of σ_A and σ_B, which can be calculated per Equation (4.43), three different versions of limit state functions can be established for a component of brittle material in a plane stress condition.

When $\sigma_A \geq \sigma_B \geq 0$,

$$g\left(S_{ut}, \sigma_A\right) = S_{ut} - \sigma_A = \begin{cases} > 0 & \text{Safe} \\ = 0 & \text{Limit state} \\ < 0 & \text{Failure.} \end{cases} \tag{4.50}$$

When $\sigma_A \geq 0 \geq \sigma_B$,

$$g\left(S_{ut}, S_{uc}, \sigma_A, \sigma_b\right) = 1 - \left(\frac{\sigma_A}{S_{ut}} + \frac{|\sigma_B|}{S_{uc}}\right) = \begin{cases} > 0 & \text{Safe} \\ = 0 & \text{Limit state} \\ < 0 & \text{Failure.} \end{cases} \tag{4.51}$$

When $0 \geq \sigma_A \geq \sigma_B$,

$$g\left(S_{uc}, \sigma_B\right) = S_{uc} - |\sigma_B| = \begin{cases} > 0 & \text{Safe} \\ = 0 & \text{Limit state} \\ < 0 & \text{Failure.} \end{cases} \tag{4.52}$$

The limit state functions (4.48)–(4.52) can be used to calculate the reliability of a component of brittle material under combined stresses due to static loadings.

We will use two examples to demonstrate how to calculate the reliability of a component of brittle material under combined stresses.

Example 4.23

A schematic of the cross-section of a column is subjected to a compressive force and bending moment as shown in Figure 4.21a, where the tensile force F_x is along the x-axis and through the centroid of the cross-section, and the bending moment M_z is about the neutral z-axis. The tensile force is $F_x = 800 \pm 60$ (lb). The bending moment is $M_z = 4000 \pm 320$ (lb/in). The cross-section of the column is shown in Figure 4.21b, where $L = 1.25 \pm 0.010''$ and $t = 0.25 \pm 0.010''$. The column is made of brittle material. Its ultimate tensile strength S_{ut} follows a normal distribution with a mean $\mu_{S_{ut}} = 22,000$ (psi) and standard deviation $\sigma_{S_{ut}} = 1800$ (psi). The ultimate compression strength S_{uc} follows a normal distribution with a mean $\mu_{S_{uc}} = 82,000$ (psi) and a standard deviation $\sigma_{S_{uc}} = 10,500$ (psi). Calculate the reliability of the component by using the MNS theory.

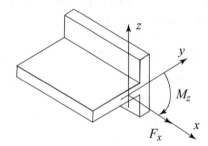   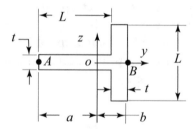

(a) Schematic of Axial Loading and Bending Moment   (b) Schematic of the Cross-section and Neutral Axis Z

Figure 4.21: Schematic of a segment of a column under compressions and bending.

Solution:

1. The maximum tensile normal stress and normal compressive stress.

   The centroid of the cross-section, as shown in Figure 4.21b, will be:

$$a = \frac{3L + t}{4}, \qquad b = \frac{3t + L}{4}. \qquad (a)$$

   The area A and the moment of inertia I of the cross-section are:

$$A = 2Lt \qquad (b)$$

$$I = Lt\left(b - \frac{t}{2}\right)^2 + \frac{Lt^3}{12} + Lt\left(a - \frac{L}{2}\right)^2 + \frac{tL^3}{12} = \frac{5Lt^3 + 5tL^3 + 6t^2L^2}{24}. \qquad (c)$$

   The critical points on the cross-section in this example will be the points on the outer layers, that is, point B and point A.

On point B, the normal stress will be the combined stress of the tensile stress due to the bending moment M_z and the tensile stress due to the tensile force F_x.

$$\sigma_B = \frac{M_z b}{I} + \frac{F_x}{A} = \frac{6M_z(3t+L)}{5Lt^3 + 5tL^3 + 6t^2L^2} + \frac{F_x}{2Lt}. \tag{d}$$

On point A, the normal stress will be the combined stress of the compression stress due to the bending moment M_z and the tensile stress due to the tensile force F_x

$$\sigma_A = -\frac{M_z a}{I} + \frac{F_x}{A} = -\frac{6M_z(3L+t)}{5Lt^3 + 5tL^3 + 6t^2L^2} + \frac{F_x}{2Lt}. \tag{e}$$

If we use the mean values to calculate normal stresses at point A and B, we will have $\sigma_A = -23296$ (pis) and $\sigma_B = 13568$ (psi). Therefore, the normal stress at point A will be compression stress, and the normal stress at point B will be tensile.

2. The limit state functions.

Per Equation (4.48), the limit state function of the column at the critical point B by using the maximum normal stress theory will be

$$g(S_{ut}, t, L, F_x, M_z) = S_{ut} - \frac{6M_z(3t+L)}{7Lt^3 + 7tL^3 + 6t^2L^2} - \frac{F_x}{2Lt} = \begin{cases} > 0 & \text{Safe} \\ 0 & \text{Limit state} \\ < 0 & \text{Failure.} \end{cases} \tag{f}$$

Per Equation (4.49), the limit state function of the column at the critical point A by using the maximum normal stress theory will be:

$$g(S_{uc}, t, L, F_x, M_z) = S_{uc} - \frac{6M_z(3L+t)}{7Lt^3 + 7tL^3 + 6t^2L^2} + \frac{F_x}{2Lt} = \begin{cases} > 0 & \text{Safe} \\ 0 & \text{Limit state} \\ < 0 & \text{Failure.} \end{cases} \tag{g}$$

Geometric dimensions can be treated as normal distributions. Their distribution parameters can be determined per Equation (4.1). The loading F_x and M_z can be treated as normal distributions. Their distribution parameters can be determined by Equation (4.2). Their distribution parameters are listed in Table 4.47.

Table 4.47: The distribution parameters of random variables in Equations (f) and (g)

S_{ut} (psi)		S_{uc} (psi)		t (in)		L (in)		F_x (lb)		M_z (lb/in)	
$\mu_{S_{ut}}$	$\sigma_{S_{ut}}$	$\mu_{S_{uc}}$	$\sigma_{S_{uc}}$	μ_t	σ_t	μ_L	σ_L	μ_{F_x}	σ_{F_x}	μ_{M_z}	σ_{M_z}
22,000	1,800	82,000	10,500	0.25	0.0025	1.25	0.0025	800	15	3200	60

3. Reliability of the column under combined stresses.

We can use the Monte Carlo method to calculate the reliability of this example based on two limit state functions (f) and (g). The Monte Carlo method has been discussed in Section 3.8. We can follow the procedure of the Monte Carlo method and the program flowchart in Figure 3.8 to create a MATLAB program. Since this problem is no very big and complicated, we will use the trial number $N = 1{,}598{,}400$ from Table 3.2 in Section 3.8. The estimated reliability R of this column at the critical point B is:

$$R = 0.9983.$$

The estimated reliability R of this column at the critical point A is:

$$R = 1.0000.$$

Therefore, the reliability of this complement in this example will be

$$R = 0.9983.$$

■

Example 4.24
A plane stress element of a component of a brittle material at the critical point is shown in Figure 4.22. The normal compressive stress σ_x (ksi), the normal tensile stress σ_y, and the shear stress τ_{xy} follows normal distributions. Their distribution parameters are listed in Table 4.48.

Table 4.48: The stresses in a plane stress element

σ_x (ksi)		σ_y (ksi)		τ_{xy} (ksi)	
μ_{σ_x}	σ_{σ_x}	μ_{σ_y}	σ_{σ_y}	$\mu_{\tau_{xy}}$	$\sigma_{\tau_{xy}}$
31.2	2.51	1.80	0.085	15.0	2.31

The component's ultimate tensile strength S_{ut} follows a normal distribution with a mean $\mu_{S_{ut}} = 22.00$ (ksi) and standard deviation $\sigma_{S_{ut}} = 1.80$ (ksi). Its ultimate compression strength S_{uc} follows a normal distribution with a mean $\mu_{S_{uc}} = 82.00$ (ksi) and standard deviation $\sigma_{S_{uc}} = 10.50$ (ksi). Calculate the reliability of the component by using the BCM theory.

Solution:

1. The two principal stresses in a plane stress.

As shown in Figure 4.22, σ_x is normal compressive stress, σ_y is normal tensile stress, and τ_{xy} is negative shear stress. In the following calculations, σ_y, σ_x, and τ_{xy} will be the values

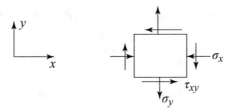

Figure 4.22: Schematic of a plane stress element at the critical point.

of these stresses. A negative sign will identify the negative stress before the stress. For this plane stress case, σ_A and σ_B can be calculated per Equation (4.42):

$$\sigma_A = \frac{-\sigma_x + \sigma_y}{2} + \sqrt{\left(\frac{-\sigma_x - \sigma_y}{2}\right)^2 + \left(-\tau_{xy}\right)^2}$$

$$\sigma_B = \frac{-\sigma_x + \sigma_y}{2} - \sqrt{\left(\frac{-\sigma_x - \sigma_y}{2}\right)^2 + \left(-\tau_{xy}\right)^2}. \tag{a}$$

From Equation (a), it is clear that $\sigma_A > 0$ and $\sigma_B < 0$. For example, we plug the mean values of σ_x, σ_y, and τ_{xy} in Equation (a) and have: $\sigma_A = 7.625$ (ksi) and $\sigma_B = -36.825$ (ksi).

2. The limit state function of this example.

   Since $\sigma_A > 0 > \sigma_B$ in this example, the limit state function of this component by using the BCM theory per Equation (4.50) is:

$$g\left(S_{ut}, S_{uc}, \sigma_x, \sigma_y, \tau_{xy}\right)$$

$$= 1 - \left(\frac{\frac{-\sigma_x + \sigma_y}{2} + \sqrt{\left(\frac{-\sigma_x - \sigma_y}{2}\right)^2 + \left(-\tau_{xy}\right)^2}}{S_{ut}} + \frac{-\frac{-\sigma_x + \sigma_y}{2} + \sqrt{\left(\frac{-\sigma_x - \sigma_y}{2}\right)^2 + \left(-\tau_{xy}\right)^2}}{S_{uc}}\right)$$

$$= \begin{cases} > 0 & \text{Safe} \\ 0 & \text{Limit state} \\ < 0 & \text{Failure.} \end{cases} \tag{b}$$

There are five random variables in the limit state function (b). All these random variables are normal distributions, and their distribution parameters are provided and displayed in Table 4.49.

3. Reliability of the component in this example.

   We can use the Monte Carlo method to calculate the reliability of this example based on the limit state function (b). We can follow the Monte Carlo method in Section 3.8 and

Table 4.49: The distribution parameters of random variables in Equations (b)

S_{ut} (ksi)		S_{uc} (ksi)		σ_x (ksi)		σ_y (ksi)		τ_{xy} (ksi)	
$\mu_{S_{ut}}$	$\sigma_{S_{ut}}$	$\mu_{S_{uc}}$	$\sigma_{S_{uc}}$	μ_{σ_x}	σ_{σ_x}	μ_{σ_y}	σ_{σ_y}	$\mu_{\tau_{xy}}$	$\sigma_{\tau_{xy}}$
22.00	1.80	82.00	10.50	31.2	2.51	1.80	0.085	15.0	2.31

the program flowchart in Figure 3.8 to create a MATLAB program. Since this problem is no very big and complicated, we will use the trial number $N = 1{,}598{,}400$ from Table 3.2 in Section 3.8. The estimated reliability R of this component at the critical point is:

$$R = 0.9417.$$

■

4.11   SUMMARY

In reliability-based mechanical design, all design parameters, including geometric dimensions, loadings, and material strengths, are treated as random variables. These statements are true in reality but require most information for their descriptions. Reliability links all design parameters through a limit state function and is a measure of components' safety status. The physical meaning of reliability is a relative percentage of safe components in the sample space of the whole same component.

When loading conditions, geometric dimensions, and the type of material are fully specified, we can calculate its reliability. Failure theories under static loadings in mechanics of materials and the reliability-based design are the same. These failure theories discussed in the traditional design theory can be used to build the limit state function. After the limit state function is established, the methods discussed in Chapter 3 including the H-L method, R-F method, and Monte Carlo method can be used to calculate the reliability of a component under any design case. The following is the summary of typical limit state functions.

For a strength issue, the typical limit state functions of a component under simple typical loading are summarized as follows.

- For a rod of brittle material under axial loading,

$$g\left(S_u, K_t, A, F_a\right) = S_u - K_t\frac{F_a}{A} = \begin{cases} > 0 & \text{Safe} \\ = 0 & \text{Limit state} \\ < 0 & \text{Failure.} \end{cases} \tag{4.12}$$

- For a rod of ductile material under axial loading,

$$g\left(S_y, K_t, A, F_a\right) = S_y - K_t \frac{F_a}{A} = \begin{cases} > 0 & \text{Safe} \\ = 0 & \text{Limit state} \\ < 0 & \text{Failure.} \end{cases} \tag{4.13}$$

- For a component of ductile material under direct shearing,

$$g\left(S_{sy}, A, V\right) = S_{sy} - \frac{V}{A} = \begin{cases} > 0 & \text{Safe} \\ = 0 & \text{Limit state} \\ < 0 & \text{Failure.} \end{cases} \tag{4.22}$$

- For a component of brittle material under direct shearing,

$$g\left(S_{su}, A, V\right) = S_{su} - \frac{V}{A} = \begin{cases} > 0 & \text{Safe} \\ = 0 & \text{Limit state} \\ < 0 & \text{Failure.} \end{cases} \tag{4.23}$$

- For a solid round shaft under torque,

$$g\left(S_{sy}, K_s, d_o, T\right) = S_{sy} - K_s \frac{16T}{\pi d_o^3} = \begin{cases} > 0 & \text{Safe} \\ = 0 & \text{Limit state} \\ < 0 & \text{Failure.} \end{cases} \tag{4.27}$$

- For a hollow round shaft under torque,

$$g\left(S_{sy}, K_s, d_o, d_i, T\right) = S_{sy} - K_s \frac{16T d_o}{\pi \left(d_o^4 - d_i^4\right)} = \begin{cases} > 0 & \text{Safe} \\ = 0 & \text{Limit state} \\ < 0 & \text{Failure.} \end{cases} \tag{4.28}$$

- For a beam of brittle material under bending:

$$g\left(S_u, K_t, Z, M\right) = S_u - K_t \frac{M}{Z} = \begin{cases} > 0 & \text{Safe} \\ = 0 & \text{Limit state} \\ < 0 & \text{Failure.} \end{cases} \tag{4.35}$$

- For a beam of ductile material under bending,

$$g\left(S_y, K_t, Z, M\right) = S_y - K_t \frac{M}{Z} = \begin{cases} > 0 & \text{Safe} \\ = 0 & \text{Limit state} \\ < 0 & \text{Failure.} \end{cases} \tag{4.36}$$

For a strength issue, the typical limit state functions of a component under combined stressed are summarized as follows.

- For ductile material, the limit state function of a component by the MSS theory,

$$g\left(S_y, \tau_{max}\right) = \frac{S_y}{2} - \tau_{max} = \begin{cases} > 0 & \text{Safe} \\ = 0 & \text{Limit state} \\ < 0 & \text{Failure.} \end{cases} \tag{4.41}$$

- For ductile material, the limit state function of a component by the DE theory,

$$g\left(S_y, \sigma_{von}\right) = S_y - \sigma_{von} = \begin{cases} > 0 & \text{Safe} \\ = 0 & \text{Limit state} \\ < 0 & \text{Failure.} \end{cases} \tag{4.45}$$

- For brittle material, the limit state function of a component by the MNS theory, when $\sigma_1 \geq \sigma_2 \geq \sigma_3 \geq 0$,

$$g\left(S_{ut}, \sigma_1\right) = S_{ut} - \sigma_1 = \begin{cases} > 0 & \text{Safe} \\ = 0 & \text{Limit state} \\ < 0 & \text{Failure;} \end{cases} \tag{4.48}$$

when $0 \geq \sigma_1 \geq \sigma_2 \geq \sigma_3$,

$$g\left(S_{uc}, \sigma_3\right) = S_{ut} - |\sigma_3| = \begin{cases} > 0 & \text{Safe} \\ = 0 & \text{Limit state} \\ < 0 & \text{Failure.} \end{cases} \tag{4.49}$$

- For brittle materials, the limit state function of a component under plane stress by the BCM theory, when $\sigma_A \geq \sigma_B \geq 0$,

$$g\left(S_{ut}, \sigma_A\right) = S_{ut} - \sigma_A = \begin{cases} > 0 & \text{Safe} \\ = 0 & \text{Limit state} \\ < 0 & \text{Failure;} \end{cases} \tag{4.50}$$

when $\sigma_A \geq 0 \geq \sigma_B$,

$$g\left(S_{ut}, S_{uc}, \sigma_A, \sigma_b\right) = 1 - \left(\frac{\sigma_A}{S_{ut}} + \frac{|\sigma_B|}{S_{uc}}\right) = \begin{cases} > 0 & \text{Safe} \\ = 0 & \text{Limit state} \\ < 0 & \text{Failure;} \end{cases} \tag{4.51}$$

when $0 \geq \sigma_A \geq \sigma_B$,

$$g\left(S_{uc}, \sigma_B\right) = S_{uc} - |\sigma_B| = \begin{cases} > 0 & \text{Safe} \\ = 0 & \text{Limit state} \\ < 0 & \text{Failure.} \end{cases} \tag{4.52}$$

For a deformation issue, the typical limit state function of a component under simple typical loading are summarized as follows.

- For a rod under axial loadings,

$$g\left(E, A_i, L, F_{ai}\right) = \Delta - \sum \frac{F_{ai}L_i}{EA_i} = \begin{cases} > 0 & \text{Safe} \\ = 0 & \text{Limit state} \\ < 0 & \text{Failure.} \end{cases} \tag{4.20}$$

- For a shaft under torque,

$$g\left(G, L_i, J_i, T_i\right) = \Delta - \sum \frac{T_i L_i}{GJ_i} = \begin{cases} > 0 & \text{Safe} \\ = 0 & \text{Limit state} \\ < 0 & \text{Failure.} \end{cases} \tag{4.33}$$

- For a beam under bending and lateral force

$$g\left(y_{\max}\right) = \Delta - y_{\max} = \begin{cases} > 0 & \text{Safe} \\ = 0 & \text{Limit state} \\ < 0 & \text{Failure.} \end{cases} \tag{4.38}$$

- For a simple support beam under a concentrated load in the middle,

$$g\left(E, I, L, P\right) = \Delta - y_{\max} = \Delta - \frac{PL^3}{48EI}. \tag{4.39}$$

- For a cantilever beam under a concentrated load on the free end,

$$g\left(E, I, L, P\right) = \Delta - y_{\max} = \Delta - \frac{PL^3}{3EI}. \tag{4.40}$$

When the limit state function of a component under static loading is established, we can use the H-L method, R-F method, or Monte Carlo method to calculate its reliability.

4.12   REFERENCES

[1] Oberg, E., Jones, F. D., Horton, K. L., and Ryffel, H. H., *Machinery's Handbook*, 30th ed., South Norwalk, Industrial Press, Incorporated, 2016. 161

[2] Callister, W. D. and Rethwisch, D. R., *Materials Science and Engineering: An Introduction*, 9th ed., Joseph Wiley, Hoboken, NJ, 2014. 168

[3] Hu, Z. and Le, Xiaobin, *Probabilistic Design Method of Mechanical Components*, Shanghai Jiao Tong University Publisher, Shanghai, China, September 1995. 168

[4] Haugen, E. B., *Probabilistic Mechanical Design*, John Wiley & Sons, Inc., 1980. 168, 174, 175

[5] Budynas, R. G. and Nisbett, J. K., *Shigley's Mechanical Engineering Design*, 10th ed., Mc-Graw Hill Education, New York, 2014. 215, 222

[6] Rao, S. S., *Reliability Engineering*, Pearson, 2015. 176, 177

[7] Pilkey, W. D., *Formulas for Stress, Strain, and Structural Matrices*, 2nd ed., John Wiley & Sons, Inc., Hoboken, NJ, 2005. DOI: 10.1002/9780470172681. 177

4.13   EXERCISES

4.1.   The length of a component is $L = 3.25 \pm 0.010''$. Determine its mean and standard deviation if it is treated as a normal distribution.

4.2.   The cross-section of a rectangular shape is with a height $h = 1.25 \pm 0.008''$ and a width $b = 2.25 \pm 0.010''$. These dimensions can be treated as normal distributions. Determine their distribution parameters.

4.3.   The concentrated load P on a beam is $P = 1520 \pm 200$ (lb). Determine its distribution parameters if it treated as a normal distribution.

4.4.   The torque T of a shaft is a uniform distribution between 2100 (lb/in) and 2500 (lb/in). Determine its PDF and distribution function.

4.5.   The bending moment M on the free end of a cantilever beam is $M = 2215 \pm 300$ (lb/in). Determine its mean and standard deviation if it is treated as a normal distribution.

4.6.   Conduct literature research to find the distribution parameters of yield strength or ultimate strength of two steel materials. The source, test method, and sample size should be included in the summary.

4.7. Conduct literature research to find distribution parameters of yield strength or ultimate strength of two aluminum alloys. The source, test method, and sample size should be included in the summary.

4.8. Conduct literature research to display distribution parameters of Young's modulus of steel. The source, test method, and sample size should be included in the summary.

4.9. Conduct literature research to display the distribution parameter of Young's modulus of an aluminum. The source, test method, and sample size should be included in the summary.

4.10. The ultimate strength S_u of material follows a normal distribution with a mean $\mu_{S_u} = 36.4$ (ksi) and a standard deviation $\sigma_{S_u} = 2.79$ (ksi). Estimate distribution parameters of its ultimate shear strength.

4.11. The yield strength S_y of a material can be described by a normal distribution with a mean $\mu_{S_y} = 68.3$ (ksi) and a standard deviation $\sigma_{S_u} = 7.12$ (ksi). Estimate distribution parameters of its shear yield strength.

4.12. The information of material shows that its yield strength will be 45.89–62.67 ksi. Estimate its distribution parameters if it is treated as a normal distribution.

4.13. The table for a material shows that its ultimate strength is between 25.67 and 35.24 ksi. Estimate its distribution parameters if it is treated as a normal distribution.

4.14. The stress concentration factor K_s of a stepped shaft under torsion is 2.17 from a design handbook table. Estimate its distribution parameters if it can be simplified as a normal distribution.

4.15. The stress concentration factor K_t of a stepped plate under bending is 1.78 from a table. Estimate its distribution parameters if it follows a normal distribution.

4.16. From three sample tests, averages of the ultimate strength, yield strength, and Young's modulus are 19.8 ksi, 24.5 ksi, and 2.45×10^4 (ksi). If its mechanical properties are treated as normal distributions, estimate distribution parameters of yield strength, shear yield strength, ultimate strength, ultimate shear strength, Young's modulus, and shear Young's modulus.

4.17. A two-bar supporter as shown in Figure 4.23 is subjected to a concentrated force F_C at the joint C. The concentrated force is $F_C = 2.1 \pm 0.20$ (klb). The bar AC and BC are pinned to a wall. The bar AC and BC have the same diameter $d = 0.25 \pm 0.005''$. The angle between the two bars is $\theta = 60 \pm 1°$. The yield strength S_y of the bar of a ductile material follows a normal distribution with a mean $\mu_{S_y} = 34.5$ (ksi) and a standard deviation $\mu_{S_y} = 3.12$ (ksi). Calculate the reliability of the bar AC. (Note: the angle needs to be converted into radian.)

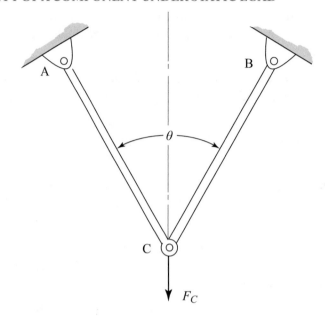

Figure 4.23: A two-bar supporter.

4.18. A two-bar supporter as shown in Figure 4.23 is subjected to a concentrated force F_C at the joint C. The concentrated force F_C can be described by a PMF:

$$p(F_C) = \begin{cases} 0.8 & \text{when } F_C = 2.0 \, (\text{klb}) \\ 0.2 & \text{when } F_C = 2.5 \, (\text{klb}). \end{cases}$$

The bar AC and BC are pinned to a wall. The bar AC and BC have the same diameter $d = 0.375 \pm 0.005''$. The angle between the two bars is $\theta = 50 \pm 1°$. The ultimate strength S_u of the bar of a brittle material follows a normal distribution with a mean $\mu_{S_u} = 82.3$ (ksi) and a standard deviation $\sigma_{S_u} = 6.32$ (ksi). Calculate the reliability of the bar AC. (Notes: the angle needs to be converted into radian.)

4.19. A bar connected to the supporter at section A is subjected to two concentrated loads F_B and F_C, as shown in Figure 4.24. The axial loads are: $F_B = 1500 \pm 120$ (lb) and $F_C = 1000 \pm 90$ (lb). The geometric dimensions are: diameter $d = 0.375 \pm 0.005''$, length of AB segment $L_1 = 8.00 \pm 0.003''$, and the length of BC segment $L_1 = 10,00 \pm 0.003''$. The Young's modulus E of the bar material follows a normal distribution with a mean $\mu_E = 2.73 \times 10^7$ (psi) and a standard deviation $\sigma_E = 1.30 \times 10^6$ (psi). Calculate the reliability of this bar when the maximum allowable deflection of the bar is $0.008''$.

4.20. A bar, as shown in Figure 4.24, is subjected to two axial loads. The axial loads are: $F_B = 2.40 \pm 0.21$ (klb) and $F_C = 1.10 \pm 0.08$ (klb). The diameter of the bar is $d =$

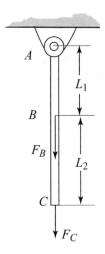

Figure 4.24: A bar under two axial loadings.

$0.25 \pm 0.005''$. The ultimate strength S_u of the bar of a brittle material follows a normal distribution with a mean $\mu_{S_u} = 82.3$ (ksi) and a standard deviation $\sigma_{S_u} = 6.32$ (ksi). Calculate the reliability of the bar ABC.

4.21. The pin at point B, as shown in Figure 4.23, is a double shear pin. The concentrated load F_C at the joint C is $F_C = 2200 \pm 300$ (lb). The diameter of the pin is $d = 3/16 \pm 0.005''$. The angle between the two bars is $\theta = 60 \pm 1°$. The shear yield strength S_{sy} of the pin of a ductile material follows a normal distribution with a mean $\mu_{S_{sy}} = 32{,}200$ (psi) and a standard deviation $\sigma_{S_{sy}} = 3630$ (psi). Calculate the reliability of the double-shear pin at point B. (Note: the angle needs to be converted into radian.)

4.22. The pin at point A, as shown in Figure 4.24, is a single shear pin. The diameter of the pin is $d = 0.250 \pm 0.005''$. The axial loads are: $F_B = 0.600 \pm 0.08$ (klb) and $F_C = 0.400 \pm 0.04$ (klb). The ultimate shear strength S_{su} (ksi) of the pin of a ductile material follows a lognormal distribution with a log mean $\mu_{\ln S_{su}} = 3.25$ and a log standard deviation $\sigma_{\ln S_{su}} = 0.181$. Calculate the reliability of the single-shear pin at point A.

4.23. A constant cross-section shaft with a diameter $d = 0.875 \pm 0.005''$ is subjected to a torque T. The torque T (lb/in) can be described by a lognormal distribution with a log mean $\mu_{\ln T} = 7.76$ and a log standard deviation $\sigma_{\ln T} = 0.194$. The shear yield strength S_{sy} of the shaft of a ductile material follows a normal distribution with a mean $\mu_{S_{sy}} = 32200$ (psi) and a standard deviation $\sigma_{S_{sy}} = 3630$ (psi). Calculate the reliability of the shaft.

4.24. Schematic of a segment of a shaft at its critical cross-section as shown in Figure 4.25 is subjected to a torque $T = 1350 \pm 95$ (lb/in). The smaller diameter d_1, the fillet radius

r and the larger diameter d_2 are: $d_1 = 0.750 \pm 0.005''$, $r = 1/32''$, and $d_2 = 1.000 \pm 0.005''$. The shear yield strength S_{sy} of the shaft of a ductile material follows a normal distribution with a mean $\mu_{S_{sy}} = 32,200$ (psi) and a standard deviation $\sigma_{S_{sy}} = 3630$ (psi). Calculate the reliability of the shaft.

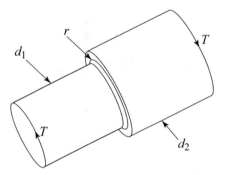

Figure 4.25: Schematic of a segment of a shaft.

4.25. A constant cross-section shaft with a diameter $d = 1.125 \pm 0.005''$ and a length $L = 15.00 \pm 0.032''$ is subjected to torque at both ends $T = 1800 \pm 120$ (lb/in). The shear Young's modulus follows a normal distribution with a mean $\mu_G = 1.117 \times 10^7$ (psi) and a standard deviation $\sigma_G = 2.793 \times 10^5$. If the design requirement is the angle of twist between two ends is less than $1°$, calculate the reliability of the shaft.

4.26. A simple support beam is subjected to a concentrated force in the middle, as shown in Figure 4.26. The concentrated force is $P = 1500 \pm 180$ (lb). The span of the beam is $L = 22 \pm 0.065''$. The diameter of this beam is $d = 1.25 \pm 0.010''$. The yield strength S_y of the beam's material follows a normal distribution with a mean $\mu_{S_y} = 34,500$ (psi) and a standard deviation $\sigma_{S_y} = 3120$ (psi). Calculate the reliability of the beam.

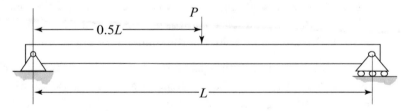

Figure 4.26: A simple support beam.

4.27. A simple support beam, as shown in Figure 4.27, is subjected to a uniformly distributed load and a concentrated load in the middle of the beam. The span of the beam is $L = 15.00 \pm 0.032''$. The beam has a rectangular shape with a height $h = 1.50 \pm 0.010''$ and

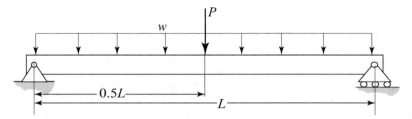

Figure 4.27: A simple support beam.

a width $b = 1.00 \pm 0.010''$. The uniformly distributed load is $w = 100 \pm 10$ (lb/in). The concentrated force in the middle is $P = 1500 \pm 180$ (lb). The yield strength S_y of the beam's material follows a normal distribution with a mean $\mu_{S_y} = 34{,}500$ (psi) and a standard deviation $\sigma_{S_y} = 3120$ (psi). Calculate the reliability of the beam.

4.28. A cantilever beam as shown in Figure 4.28 is subjected to a concentrated force at the free end $P = 150 \pm 80$ (lb). The cross-section of the beam is a rectangular shape with a height $h = 2.00 \pm 0.010''$ and a width $b = 1.00 \pm 0.010''$. The length of the beam is $L = 20.0 \pm 0.032''$. The Young's modulus of the beam material follows a normal distribution with a mean $\mu_E = 2.76 \times 10^7$ (psi) and a standard deviation $\sigma_E = 6.89 \times 10^5$ (psi). If the maximum allowable deflection of the beam is $\Delta = 0.022''$. Calculate the reliability of this beam for a deformation issue.

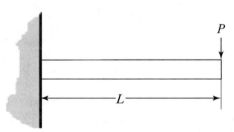

Figure 4.28: A cantilever beam.

4.29. A shaft with a diameter $d = 1.500 \pm 0.005''$ is subjected to a torque $T = 3000 \pm 150$ (lb/in) and a bending moment $M = 9000 \pm 600$ (lb/in). The yield strength S_y of the beam's material follows a normal distribution with a mean $\mu_{S_y} = 34{,}500$ (psi) and a standard deviation $\sigma_{S_y} = 3120$ (psi).

(a) Calculate the reliability of the shaft by using the MSS stress theory.

(b) Calculate the reliability of the shaft by using DE theory.

4.30. A thin-wall cylindrical vessel has a wall thickness $t = 0.25 + 0.020''$ and an inner diameter $d = 40.0 \pm 0.125''$. The internal pressure of the fluid is $p = 350 \pm 50$ (psi). The yield strength S_y of the beam's material follows a normal distribution with a mean $\mu_{S_y} = 34{,}500$ (psi) and a standard deviation $\sigma_{S_y} = 3120$ (psi).

(a) Calculate the reliability of the shaft by using the MSS stress theory.

(b) Calculate the reliability of the shaft by using DE theory.

4.31. A plane stress element of a component of a ductile material at the critical point is shown in Figure 4.29. The normal compressive stress σ_x (ksi), the normal tensile stress σ_y, and the shear stress τ_{xy} follows normal distributions. Their distribution parameters are listed in Table 4.50.

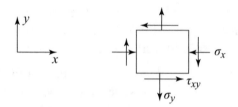

Figure 4.29: Schematic of a plane stress element at the critical point.

Table 4.50: The stresses in a plane stress element

σ_x (ksi)		σ_y (ksi)		τ_{xy} (ksi)	
μ_{σ_x}	σ_{σ_x}	μ_{σ_y}	σ_{σ_y}	$\mu_{\tau_{xy}}$	$\sigma_{\tau_{xy}}$
16,200	1,550	6,800	540	13,000	1100

The column is made of a ductile material. The yield strength S_y of the beam's material follows a normal distribution with a mean $\mu_{S_y} = 34{,}500$ (psi) and a standard deviation $\sigma_{S_y} = 3120$ (psi). Calculate the reliability of the shaft by using the DE theory.

4.32. Schematic of a critical cross-section of a column is shown in Figure 4.30. It is subjected to a compression force $F_x = 4000 \pm 180$ (lb) and a bending moment $M_z = 28{,}000 \pm 300$ (lb/in). The geometric dimensions of the critical cross-section of the column are $h = 2.25 \pm 0.010''$ and $b = 1.00 \pm 0.010''$. Its ultimate tensile strength S_{ut} follows a normal distribution with a mean $\mu_{S_{ut}} = 22.00$ (ksi) and a standard deviation $\sigma_{S_{ut}} = 1.80$ (ksi). The ultimate compression strength S_{uc} follows a normal distribution with a mean $\mu_{S_{uc}} = 82.00$ (ksi) and a standard deviation $\sigma_{S_{uc}} = 10.50$ (ksi). Calculate the reliability of the column by using the MNS theory.

4.33. A plane stress element of a component of a brittle material at the critical point is shown in Figure 4.31. The normal tensile stress σ_x (ksi), the normal compression stress σ_y and

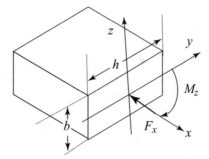

Figure 4.30: Schematic of compression force and a bending moment.

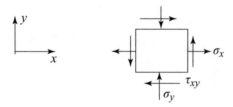

Figure 4.31: Schematic of a plane stress element.

the shear stress τ_{xy} follows normal distributions. Their distribution parameters are listed in Table 4.51.

Table 4.51: The stresses in a plane stress element

σ_x (ksi)		σ_y (ksi)		τ_{xy} (ksi)	
μ_{σ_x}	σ_{σ_x}	μ_{σ_y}	σ_{σ_y}	$\mu_{\tau_{xy}}$	$\sigma_{\tau_{xy}}$
14.2	1.51	50.80	6.25	15.0	2.31

The column is made of brittle material. Its ultimate tensile strength S_{ut} follows a normal distribution with a mean $\mu_{S_{ut}} = 22.00$ (ksi) and standard deviation $\sigma_{S_{ut}} = 1.80$ (ksi). The ultimate compression strength S_{uc} follows a normal distribution with a mean $\mu_{S_{uc}} = 82.00$ (ksi) and standard deviation $\sigma_{S_{uc}} = 10.50$ (ksi). Calculate the reliability of this component by using the BCM theory.

APPENDIX A

Samples of MATLAB® Programs

A.1 THE H-L METHOD FOR EXAMPLE 3.11

```
% The H-L method for exmaple 3.11
% The Limit State function: g(Sy,Z,P,L)=Sy-P*L/(4Z)
% Input the distribution parameters mx-mean, sz-standard deviation
clear
mx=[6*10^5,10^(-4),10,8];              %The mean
sx=[10^5,2*10^(-5),2,2.083*10^(-2)]; % The standard deviation
beta=0; % Set beta=0
% Pick an initial design point x0(i)
for i=1:3
   x0(i)=mx(i);  % Use the mean for the first n-1 variable
end
% The last one x0(4) will be determined by using the limit state function
x0(4)=4*x0(1)*x0(2)/x0(3);
% Initial point in standard normal distribution space
for i=1:4
   z0(i)=(x0(i)-mx(i))/sx(i);
end
% Start iterative process
for j=1:1000
% The Tylor series coeffcient
G0(1)=sx(1)*1;
G0(2)=sx(2)*x0(3)*x0(4)/4/(x0(2))^2;
G0(3)=sx(3)*(-1)*x0(4)/4/x0(2);
G0(4)=sx(4)*(-1)*x0(3)/4/x0(2);
% Calculate the reliability index beta0
g00=0;
z00=0;
for i=1:4
```

```
g00=g00+G0(i)^2;
z00=z00+(-1)*z0(i)*G0(i);
end
Gi0=g00^0.5;
beta0=z00/Gi0;
% Store the data of iterative process
   for i=1:4
       ddp(j,i)=x0(i);
   end
ddp(j,4+1)=beta0;
ddp(j,4+2)=abs(beta0-beta);
% New design point
% The values for the first n-1 are determined by the
% recurrence equation
 for i=1:3
     z1(i)=(-1)*beta0*G0(i)/Gi0;
     x1(i)=sx(i)*z1(i)+mx(i);
 end
% The value for the last variable will be determined
% by the limit state
% function
   x1(4)=4*x1(1)*x1(2)/x1(3);
   z1(4)=(x1(4)-mx(4))/sx(4);
% Check the convengence condition
   if ddp(j,4+2)<=0.0001;
   break
   end
% Use new design point to replace previous design point
   for i=1:4
       z0(i)=z1(i);
       x0(i)=x1(i);
   end
   beta=beta0;
 end
% Calculate and display reliability
format short e
disp('reliability')
R=normcdf(beta0)
% Display iterative process and write it to Excel file
```

```
disp(ddp)
xlswrite('example3.11',ddp)
```

A.2   THE R-F METHOD FOR EXAMPLE 3.13

```
% The R-F method for example 3.13
% The Limit State function: g(T, d, Ssy)=Ssy*pi/10*d^3-T
% Input the distribution parameters dp1 and dp2
clear
mx=[34, 2.125, 31]; %The mean or the first distribution
                    %parameter
sx=[3,0.002,2.4];   %The standard deviation or the second
                    % distribution
                    %parameter
% Calculate the mean of T (Weibull) and the initial
% point x0(i)
x0(1)=mx(1)*gamma(1/sx(1)+1);
x0(2)=mx(2);
% The value of the last variable is determined by the
% limit state function
x0(3)=16*x0(1)/pi/x0(2)^3;    % Equation (3.71) or (d)
beta=0;                       %set the beta =0
% Iterative process starting
for j=1:1000
% Calculate the equivalent mean and standard deviation
zteq=norminv(wblcdf(x0(1),mx(1),sx(1)));
steq=normpdf(zteq)/wblpdf(x0(1),mx(1),sx(1));
mteq=x0(1)-zteq*steq;
% Mean and standard deviation matrix
meq(1)=mteq;
seq(1)=steq;
for i=2:3
   meq(i)=mx(i);
   seq(i)=sx(i);
end
% Calculate z0(i)in the standrad normal distribution space
for i=1:3
   z0(i)=(x0(i)-meq(i))/seq(i);
end
```

```
% Calculate the Taylor Series Coefficient, Equation (g)
Gi(1)=-seq(1);
Gi(2)=seq(2)*x0(1)*3*pi*(x0(2))^2/16;
Gi(3)=seq(3)*pi*x0(2)^3/16;
g00=0;
z00=0;
for i=1:3
g00=g00+Gi(i)^2;
z00=z00+(-1)*z0(i)*Gi(i);
end
Gi0=g00^0.5;
% Calculate the reliability index beta0
beta0=z00/Gi0;
% Data of iterative process
for i=1:3
    ddp(j,i)=x0(i);
end
ddp(j,3+1)=beta0;
ddp(j,3+2)=abs(beta0-beta);
% new design point
 for i=1:3-1
     z1(i)=(-1)*beta0*Gi(i)/Gi0;
     x1(i)=seq(i)*z1(i)+meq(i);
 end
 x1(3)=16*x1(1)/pi/x1(2)^3;
 z1(3)=(x1(3)-meq(3))/seq(3);
% Check the convengence condition
  if ddp(j,3+2)<=0.0001;
  break
  end
% use new design point to replace previous design point
  for i=1:3
     x0(i)=x1(i);
  end
  beta=beta0;
end
% calculate and display reliability
format short e
disp('reliability')
```

```
R=normcdf(beta0)
% Display iterative process and write it to Excel file
disp(ddp)
xlswrite('example3.13',ddp)
```

A.3   THE MONTE CARLO METHOD FOR EXAMPLE 3.14

```
% Monte Carlo Method for Example 3.14
% Limit State function: g(Sy,P1,P2,d)
% =Sy-(4P1)/(?d^2 )-(4P2)/(?d^2 )
clear
% Input the data
msy=61.5;% mean of the yield strength
ssy=5.95;% standard deviation of the yield strength
mp1=10.2;% mean of the force P1
sp1=1.2;%standard deviation of the force P1
cp2=4.5;%the scale parameter of the force P2
sp2=1.5;%the shape function of the force P2
md=0.75;% mean of the diameter
sd=0.003;% standard deviation of diameter
% The trial number N
N=1598400;
% Generate the sample data for each random variable
Rsy=random('norm',msy,ssy,1,N);     % the sampling for yiled strength
Rp1=random('norm',mp1,sp1,1, N);    % Random sampling for force p1
Rp2=random('wbl',cp2,sp2,1,N);      % Random sampling for force p2
Rd=random('norm',md,sd,1,N);        % Random sampling for dimension d
% Start the Monte Carlo Method
for J=1:N
    gJ=Rsy(J)-4*Rp1(J)/pi/Rd(J)^2-4*Rp2(J)/pi/Rd(J)^2;
    if gJ >= 0
       NT(J)=1;
    else
       NT(J)=0;
    end
end
% Calculate the reliability
RR=0
for J=1:N
```

```
    RR=RR+NT(J);
end
display ('Reliability R')
R=RR/N
display ('The relative error')
rerror=2*(R/N/(1-R))^0.5
```

Author's Biography

XIAOBIN LE

Xiaobin Le, Ph.D., P.E., received a BS in Mechanical Engineering in 1982 and an MS in Mechanical Engineering in 1987 from Jiangxi University of Science and Technology, Ganzhou, Jiangxi. He received his first Ph.D. in Mechanical Design of Mechanical Engineering from Shanghai Jiao Tong University, Shanghai, in 1993, and his second Ph.D. in Solid Mechanics of Mechanical Engineering from Texas Tech University, Lubbock, Texas, in 2002.

He is currently a professor in the Mechanical Engineering Department at Wentworth Institute of Technology, Boston, Massachusetts. His teaching and research interests are Computer-Aided Design, Mechanical Design, Finite Element Analysis, Fatigue Design, Solid Mechanics, Engineering Reliability, and Engineering Education Research.

Printed in the United States
by Baker & Taylor Publisher Services